Contractors, Engineers, Architects, Inspectors & Unions
"Can Improve or Recuperate 40% of the:
Quality, safety, performance, leadership,
Innovation, competitiveness and earnings of any
Corporation, Contractor or profession."

Theory & Practical KNOWLEDGE

IMPROVE our own MANAGEMENT

SUCCESS

Adopt & Adapt New: { Changes, Information & Technology

By: **EMILIO CRISTANCHO G.; BS, MBA**
California State University
Ex-Executive Segundo Cristancho J.& Co. and
Multinational "BECHTEL" Corp., California, USA.

In Collaboration with:
Dr. Gilberto Ortiz-Gonzalez; MD
Jorge and Octavio Cristancho-G.; BS
Dr. Maria Mercedes Cristancho -O.; MD

1

We dedicate this book as a tribute:
To our parents:
**SEGUNDO CRISTANCHO J. and
LOLA G. de CRISTANCHO**
To the pleasant recollection of her and
To the noble and hard working life of his,
and for theirs excellent love and care and
for teaching us theirs:
Brilliant business wisdom and also
for their immense sacrifice to give us
the best education and wellbeing.

Also to my father in law:
Dr. GILBERTO ORTIZ GONZALEZ, MD,
For his example and teachings and for being
a second father to me; and also
To my family for their love, help and
endurance all these years.

Also thanks to all our friends for their help,
companionship and moral support, thank you;
and finally to my dear colleagues at:
Segundo Cristancho J. and Co., and the
Multinational Bechtel Corp., Thank you
for all your coaching, training, help and
friendship; It was a privilege
sharing your wisdom with me,
I wish all of you excellent:
Health, Happiness and Prosperity,
May God bless all of you !!.
Emilio Cristancho-G.

This book contains practical management wisdom,
From brilliant world renowned: Presidents (CEOs),
Professionals (Contractors, Engineers, Architects, Inspectors & Unions),
Universities, and MDs mentioned;
To help us Improve **How we Manage "Ourselves."**

It is advisable, to apply the golden rule of psychology
"BE VERY SINCERE with yourself"
When analyzing, which symptoms need to be modified,
To be able, to modify the Super-Ego* and the Emotional Intelligence.

This book will help us obtain, an
Excellent personal management style,
"To improve or regain 40% of the:
Quality, safety, performance, leadership,
Innovation, competitiveness and earnings of any
Corporation, contractor, project or profession
{Mr. H. Geneen, Ex-CEO of ITT};"*(2) and also,
To achieve professional, emotional and financial sustainable success.

According to the world famous
Dr. Emilio Mira y Lopez, MD:
*The Super-ego should be called Anti-ego, because
There is a sector, with psychological forces hostile to the Ego.*(5)pag202

*The Super-ego also defines the psychological or
Social environment,
Where we: Grew, lived, studied and **Imitated** the
Behaviors and **Values**: Ethical, moral, religious and cultural.

Warning: This book is not intended to take the place of medical advice
From a trained medical professional; thus, always seek the best medical
Assistance in ALL cases regarding your total health. Neither the authors,
Nor the publisher of this book, nor the publishers of the books mentioned
in this book take any responsibility if this warning is not followed completely.

*(2) Managing, by H. Geneen (Ex-CEO ITT), Pag (184-5) Pub. by Avon Books..
*(5) Dr. Emilo Mira y Lopez, MD: The four Giants of the Soul, pag 202, Edit. Lidium

TABLE of CONTENTS:

Contractors, Engineers, Architects and Inspectors "Can Improve or Recuperate 40% of the: Quality, safety, performance, leadership, Innovation, competitiveness and earnings of any Corporation, contractor or profession {Mr. H. Geneen, Ex-CEO of ITT}."*(2)

Section I: How To Manage "Yourself" (Soft Skills):

A) To improve our own management (the super-ego),*
It is advisable, to apply the golden rule of psychology,
"Be very sincere with yourself," because
Knowing ourselves better, it will allow us to identify,
1) Our natural talents to be developed, and
2) Our personality's deficiencies that must be eliminated;
To be able to improve our super-ego and Emotional Intelligence; and
To listen with better: Empathy, respect and admiration.

This way, we will be able to apply the following definitions,
To improve the innovations or changes required,
For the competitiveness and survival of the:
Corporation, contractor, project or profession:
a) How to modify the personality
b) Egotism**
c) An excellent Engineer or Inspector
d) Wealth: { **Inner = 60%** &
{ **Outer = 40%**

Section II:{EMOTIONAL over Intellectual} Intelligence
Section III: Executive Summary
According to the world famous,
Dr. Emilio Mira y Lopez, MD:
"The super-ego should be called Anti-ego, because
There is a sector, of psychological forces hostile to the Ego."*(5)

The super-ego also defines the psychological = **Social environment,**
Where we: Grew, lived, studied and **Imitated** the
Behaviors and **Values:** ethical, moral, religious and cultural.
**Egotism = Arrogant, or abusive or disrespectful behavior.
*(2) Managing by: Harold Geneen, Pag 185, Pub. Avon Books.
*(5)Dr. Emilio Mira y Lopez, MD: The four Giants of the Soul, pag 202, Edit. Lidium

Section I: How To Manage "Yourself" (Soft Skills):
To be able, **"To improve or regain 40% of the:**
Quality, safety, performance, leadership,
Innovation, competitiveness and earnings of any
Corporation, contractor or profession
{Mr. H. Geneen, Ex-CEO of ITT}."*(2)

It is advisable, **To Analyze, Compare & Apply**
Sections I & II of this practical book,
Applying the wisdom of the world renowned: Presidents (CEOs),
Professionals (Contractors, Engineers, Architects, Inspectors & Unions),
Universities, and MDs mentioned.

It is advisable, to apply the golden rule of psychology:
"Be Very Sincere with "yourself";
When Analyzing, which symptoms need to be modified, to help
Modify the Super-Ego* and the Emotional Intelligence.

This way, the following definitions can be applied,
To improve the innovations or changes required,
For the competitiveness and survival of the:
Corporation, contractor, project or profession:
- **a)** How to modify the personality
- **b)** Egotism**
- **c)** An excellent Engineer or Inspector
- **d)** Wealth: $\begin{cases} \textbf{Inner = 60\%} \ \& \\ \textbf{Outer = 40\%} \end{cases}$

According to the world famous,
Dr. Emilio Mira y Lopez, MD:
*"The super-ego should be called anti-ego, because sometimes,
There is a sector, of psychological forces hostile to the Ego."*(5)pag. 202

*The super-ego also defines the psychological or
Social environment where we:
Grew, lived, studied and **Imitated** the
Behaviors and **values**: ethical, moral, religious and cultural.
*Super-ego = In psychoanalysis, is that part of the psyche which:
- **1)** Is critical of the self or Ego, and
- **2)** Enforces moral standards, at an unconscious level,
   {**If** we have modified all the deficiencies of the Super-Ego};

Then, it blocks the irrational unacceptable impulses of the **Id**.*
*The **Id** = Irrational: instincts, impulses, passions; controls the pain & pleasure.
**Egotism = Arrogant, or abusive or disrespectful Behavior.

Section I: How To Manage "Yourself" (Soft Skills):

According to world renowned management adviser
Dr. Peter Drucker: "Managers should indeed know more about
Human beings, above all, managers and
{Professionals = Contractors, Engineers, Architects, Inspectors and Unions}
Need to know much more about "Themselves" (soft skills),**
so that they will not impair performance."*(1)

In other words: **"Know Thyself,"***(1)
To be able, to improve:
How we manage **"Ourselves"**,
Before we can manage **"Others"** successfully, and
"Improve or regain **40% of the**
Quality, safety, performance, leadership,
Innovation, competitiveness and earnings of any
Corporation, contractor or profession
{Mr. H. Geneen, Ex-CEO of ITT}"*(2)
{Advisable to modify the Super-Ego* and the Emotional Intelligence (E.I.)}.

Consequently, it is advisable,
"To Be very sincere with "yourself",
When analyzing the symptoms to be modified;
To improve the super-ego* and the Emotional Intelligence (E.I.),
To be able, to achieve sustainable success, at
The Professional, emotional and financial level.

Note: On Nov. 11/2002, Business Week mentioned that:
"Some Presidents (or Professionals) are:
HIRED for their excellent Professional Knowledge and
FIRED for theirs Personalities"= Un-Modified super-ego* and E.I.

**Soft Skills = Intelligence {Emotional + Social}
Hard Skills = Knowledge {Professional, Scientific + Practical}

According to the world famous,
Dr. Emilio Mira y Lopez, MD:
The super-ego should be called anti-ego, because,
There is a sector, of psychological forces hostile to the Ego."*(5) pag.202
*The super-ego = The psychological or Social environment,
Where we: Grew, lived, studied and **Imitated** the
Behaviors and **values**: ethical, moral, religious and cultural.
**Egotism = Arrogant, or abusive or disrespectful Behavior.

*(1)Management:Tasks, Responsibilities and Practices by P. Drucker, Pag. 244, Pub. Harper & Row
**(2) Managing By: Harold Geneen, Pg 185, Pub. by Avon Books.

Cont Section I: How To Manage "Yourself" (Soft Skills):

The brilliant ex-Director of ITT, Mr. Harold Geneen, mentions in his book: Managing (Pag.185)*(2), that: if we eradicate the "Egotism"** in "some" companies {un-modified Super-Ego and Emotional Intelligence} we can:

Improve 40% in:	{ Performance, Productivity & Earnings ($$$) }	Of any Company"*(2) or Profession

It is very important for managers and professionals To Improve How to manage themselves, to obtain:
a) An excellent {Emotional + Social} Intelligence,
b) To cultivate the: Inner Peace or satisfaction;
c) To be able, to apply:
                the golden rule of psychology:
        "Be very sincere with yourself,"
To modify the super-ego* and the Emotional Intelligence,
    Applying sections I & II of this valuable book,
d) To be able, to achieve sustainable Success at:
    The Professional, Emotional and Financial Level.

Mr. Geneen also mentions a very wise advice:
"More Careers are ruined by success than by failure"
{Due to deficiencies of the super-ego*, sometimes
    the success goes way over their heads, and
Frequently they start mistreating: others or themselves}.

It is advisable, to review sections I & II of this book,
always "being very sincere with yourself," to be able to follow,
    the advice given by world renowned professionals,
        To achieve sustainable success at:
    The Professional, Emotional and Financial Level.

According to the world famous, Dr. Emilio Mira y Lopez, MD:
*"The super-ego should be called anti-ego, because there is a sector,
of psychological forces hostile to the ego."*(5)pag202.

*The super-ego = The Social or psychological environment
    where we: Grew, lived, studied and **Imitated** the
Behaviors and **values**: ethical, moral, religious and cultural.
**Egotism = Arrogant, or abusive or disrespectful Behavior.
        *(2) Managing by: Harold Geneen, Pag 185, Pub. Avon Books.

Cont Section I: How To Manage "Yourself" (Soft Skills):

Matching points (*=*) of the **Management styles** of:
*Segundo Cristancho J. & Company {during 67 years}, and
*"Mr. Harold Geneen, Ex-CEO of ITT
{Helped a $766 million telephone company grow into a
$22 Billion multinational conglomerate, according to
His book: Managing, Pub. by Avon Books}:*(2)
Several executives, that worked with Mr. Geneen,
graduated to successful careers as Fortune 500 CEOs:

*=*1) **Plan, direct and control:** "Know every facet of the businesses" and

= 2) Treat people with respect & admiration and listen with empathy,
have an open door, but avoid incompetence and neglect.

*=*3)   Urgently, If present, modify the egotism,**
To be able, "To improve or regain 40% of the:
Quality, safety, performance, productivity,
Innovation, competitiveness and earnings of any
Corporation, contractor or profession
{Mr. H. Geneen, Ex-CEO of ITT}."*(2)".
{Modify the super-ego* and the Emotional Intelligence,
applying sections I & II of our practical book}.

*=*4) "Update the plans, associates must update them too.
Use what worked, but changed when necessary."

*=*5) "Be a: {Hands on, present, active and a responsible} manager."

*=*6) "Know:{your business, what works, your people, who works}

*=*7) "Manage with respect & admiration (w/o egotism)."**
"Performance and results must be achieved,
If you don't achieve those results, you're not managing"
{To Improve your management, apply sections I & II of our book}.

*=*8) "Put in the time and dedication to work things out."
Never delegate the responsibilty of controlling a
Task or a goal, to be done, with excellent:
Quality, safety, on time and under budget.

*=*9) "When a client reports a complaint or suggestion,
Communicate it to management ASAP, and fix it asap,
according to the plans, contract and specifications,
to comply with the client wishes and desires; thus
always having a satisfied and loyal client."

*Also, applied at: **Segundo Cristancho J. & Co.** during 67 years.
*(2)Managing: Harold Geneen, ex- CEO, ITT, Pub. by Avon Books.
**Egotism = Arrogant, or abusive or disrespectful Behavior

Cont Section I: How To Manage "Yourself" (Soft Skills):
How our mind* works:
the thoughts of our mind
are converted into words,
and those words will become our
behavior, habits and
Values: Ethical, moral and religious;
which will define our personality (super-ego) and
Our success in life.

Our mind contains our:
Super-Ego = **Moral Censor,**
Social environment, or
psychological environment in which we:
Grew, lived and Developed in life
and this is where we:
Learned, copied and imitated our
examples or models of:
behavior, habits and
Values: Ethical, Moral and Religious
which will define our personality and our:
Success in life.

Consequently, please remember, that to be able
to modify the personality (the super-ego), we must:
ANALYZE, COMPARE & APPLY
everything that we read and :
**BE VERY SINCERE
WITH OURSELVES.
TO MODIFY:
THE PERSONALITY
"THE SUPER-EGO"**

It is advisable, **To Analyze, Compare & Apply**
sections I & II of this practical book,
applying the wisdom of the world renowned:
Presidents (CEOs), Professionals (Engineers, Architects & Inspectors),
Contractors, Universities, and MDs mentioned;
To achieve sustainable success at
The Professional, Emotional and Financial level.

Cont Section I: How To Manage "Yourself" (Soft Skills):

TO MODIFY The PERSONALITY (Super-Ego):
List of "possible" deficiencies of the Super-Ego,
that will be convenient to modify,
"To improve or regain 40% of the:
Quality, safety, performance, leadership,
Innovation, competitiveness and earnings of
The corporation, contractor, project or profession."*(2)

Drs. Friedman & Rosenman, MDs, Dr. Daniel Goleman, H. Geneen
and many CEOs, MDs, and scientists,
Recommend to modify the personality (character),
To apply the golden rule of psychology:
"Be very Sincere with Yourself,"
When analyzing, which of the following symptoms needs to be modified,
To improve the Super-Ego and the Emotional Intelligence:

A) **Easy to Diagnose Symptoms:**
    1) Hyper-aggressiveness or hostility or
    2) Time urgency = Hurry sickness, or
    3) "Egotism = Arrogant, or abusive or disrespectful behavior, or
               Bad temper"*(2),*(4),*(5), or
    4) Problems with the Emotional Intelligence, to be able
           To listen and treat ourselves and others better;
           To avoid causing major: mistakes, accidents, or losses.

B) Hard to Diagnose Symptoms:
    1)   **a)** Self Esteem, and/or
       Un-Modified {   Insecurity (Inner Peace) and/or
                   b) Mental & } Development
                       Material }

           To avoid the:
    2) (Unconscious) Drive toward **"self-destruction"** of the:
           a) Career or business or Marriage and/or
           b) The personality or **the life.**

*(2) Managing, by H. Geneen (Ex-CEO ITT), Pag (184-5) Pub. by Avon Books.
*(4) Drs.Friedman & Rosenman, MDs and D. Ulmer: Treating type A Behavior & Your Heart,
     Pags (84,87,93,229) Published by F. Crest, N.Y., N.Y.
*(5) Dr. Emilio Mira y Lopez, MD, The four Giants of the Soul:
     Fear, Anger, Love and Duty pags 124,5, Edit. Lidiun

Some of the problems of: Quality, Safety and productivity are caused by:

Lack of: { Knowledge or
Attention or
Care or rest,

But "Sometimes" Problems of Quality, safety and Productivity are Caused (5-10)% Due to the Following reasons: {

Personal or marital problems or
Prepotency and/or narcissism or
Problems working as a team member or
Addiction to: { Substances or
Alcohol or
Sex or Persons or
Lack of Values* or pride in the work
Performed, producing perception
Differences of: { Quality, Safety
and Productivity.

*Values = {Ethical, Moral and Religious}.

Summary:
Besides working hard and knowing their job's scope,
all contractors, Engineers and Inspectors must:
follow the advice of world re-known
management's adviser Dr. Peter Drucker:
"Managers and their associates should
know more about human beings,
above all, they all need:
to know more about "themselves,"
so that they will **not** impair performance."*(1)

In other words:
"Know & Modify Thyself"
{Modify the super-ego* and the Emotional Intelligence},
To modify the personality {character}
and achieve sustainable success at
the Professional, Emotional and Financial level.

*(1) Management Tasks, Responsibilities and Practices by:
Peter F. Drucker, Pag. 244, Pub. by Harper & Row.

PRUDENCE   IS VERY VALUABLE   !!!

- Espera tío no le pegues,

Vamos a hablar las cosas

Cont Section I: How To Manage "Yourself" (Soft Skills):
Practical Applications of Good Emotional Intelligence:
Personal and Professional Management

To perform an excellent Personal and Professional Management analysis:
It is always advisable to forecast
The three (3) probable case scenarios:
1) The Best
2) The "Medium" and
3) The Worst probable case scenario,
And the contingency plans for each scenario,
in case something goes wrong or unforeseen;
Because an excellent professional with good
Soft skills: = {Emotional + Social} Intelligence
Will be able to realize,
That any of those three (3) scenarios
Can happen at any time
The best, medium or worst case scenarios;
and this is an indication of :
Excellent Emotional Intelligence at work !!!.

Also, when analyzing contractors or projects remember:
If there is a chance that something will go wrong, it will,
Unless we take precautions to avoid or mitigate it; also
If there are 10 good positive outcomes, and 2 negative outcomes,
Frequently the 2 negative outcomes will happen first.
Thus, when performing the contingency plans, always
Utilize an excellent modified Super-ego*and Emotional Intelligence
For the three (3) case scenarios: the best, medium or worst.

Note: HOW **To Delegate** a Task or a Goal:
We can delegate a task or a goal to be achieved; **but**
We can not delegate the responsibility of controlling:
That the task or goal is being achieved with the **best:**
Quality, safety, on time and under budget; **thus,**
The controlling responsibility
Always belongs to the contractors, the professionals & the team.

Cont Section I: How To Manage "Yourself" (Soft Skills):
This definition applies to all the professions pertaining to:
Engineering & Construction:
Contractors, Engineers, Architects, Inspectors and Unions.

Definition of a **PROFESSIONAL**:
1) A true professional possesses two types of knowledge:

Types of
knowledge:
{
  1) The Professional or practical wisdom, and
  2) The **one self's knowledge** =
     Modified: { Emotional Intelligence and Super-Ego*;
}

To be able, To avoid mistakes, accidents or losses.
2)   Always applies the golden rule of psychology:
"Be very sincere with yourself,"
To modify the Emotional Intelligence and the Super-Ego* pertaining to:
"The Fear, Love, Duty and
Anger: Hatred, enviousness, prepotency, bad temper, and egotism,**
mentioned by the world famous Dr. Emilio Mira y Lopez, MD
{In his book, "The four Giants of the Soul", Edit. Lidiun}."*(5)

3) Optimizes the relationships with others, to listen to them with
Empathy, respect and admiration; to implement the
Innovations and changes required,
To improve or recuperate 40% of the
Quality, safety, performance, leadership,
Innovation, competitiveness and earnings of the:
Country, corporation, contractor, project or profession.

4) A Professional continuously improves the **Inner** wealth;
To be able, to improve or preserve the **Outer** wealth.

On Nov. 1/202, Business Week mentioned that,
"Some Professionals are:
Hired for their excellent professional knowledge and
Fired for their Personalities" = Un-modified Super-Ego*and Emotional Intelligence.

According to the world famous,
Dr. Emilio Mira y Lopez, MD, the
*Super-Ego = Anti-Ego, because there is a sector, of
psychological forces hostile to the Ego.*(5) pag 202

The super-ego is also, the psychological or **Social environmen**t,
where we: Grew, lived, studied and **Imitated** the behaviors and
Values: ethical, moral, religious and cultural.
Egotism = Arrogant, or abusive or disrespectful behaviors.
*(2) Managing, by H. Geneen, Ex-CEO ITT, Pag (184-5) Pub. by Avon Books.

Cont Section I: How To Manage "Yourself" (Soft Skills):

Practical applications of good Emotional Intelligence,
PERSONAL, PROFESSIONAL OR FINANCIAL Management:

Always remember, the mind* contains our :
Super-Ego = **Moral Censor** =
Our psychological or **Social environment,**
Where we: Grew, lived and developed in life;
and this is where we:
Learned, Copied and **Imitated**
Our examples and models of
Behaviors, Habits and
Values: Ethical, Moral and Religious,
Which will define our personality and our
Success in life.

Some of the personal, professional or financial problems,
Could have been avoided,
Implementing a modified super-ego*and Emotional Intelligence,
Which prevents:
Overconfidence, or over-complacency or arrogance,
or hatred, or hostility, or narcissism, or
Not following advice or procedures;
Leading "frequently" to major:
Mistakes or losses or
"Sometimes" producing an:
Un-conscious drive toward "Self-destruction" of the:
a) Career, Business or Marriage and/or
b) The personality or the **Life.**

Consequently, it is imperative, that
We acquire more insight into our own:
Strengths and weaknesses;
To be able, to modify our Personalities
{Modify the super-ego and the Emotional Intelligence}
To obtain **success** at:
The Professional, Emotional and Financial level.

Cont Section I: How To Manage "Yourself" (Soft Skills):

"Emotional Intelligence (EI) is a combination of:
    1) Self Management and
    2) Learning Social skills in an excellent environment,
        To transform and optimize,
        The performance of the individual or the group."*(10)
According to the world famous, Dr. Emilio Mira y López:
"Healthy mind in a healthy {body and society}"*(17)pag18
{Achievable Modifying the Super-Ego*and the Emotional Intelligence}.

*Empathy: It is how well we **LISTEN** to the point of view of
    Another person, putting ourselves in their shoes,
When solving problems working as a team member;
Thereby, implementing an excellent Leadership role.

EMOTIONS: "In Dr. Goleman's words,
"Personal" competence, comes from being
    Aware of and regulating **one's own** emotions.
"Social" competence is:
Awareness and regulation of **others'** emotions."*(8)

World famous Dr. Erich Fromm, MD,
Mentions in his book "The Art of Listening" Pages (68-9):
"Only a fundamental transformation of our
    Personality system
can produce a significant change of our **character;**
    In other words:
It is advisable, to change not only in one aspect,
But **in all our aspects of our personality system;**
    That is, the way we:
Think, act, feel, move .... and everything else
Because, one isolated Emotional change,
Never produces a lasting effect"
{Modify the super-ego* and the Emotional Intelligence}.

*(8) Dr. D. Goleman, The Brain and Emotional Intelligence:
New Insights, Published by More than Sound, 2011.
*(10) Special collections, The eading Teams with Emotional Intelligence, by
Drs: Daniel Goleman, R. Boyatsis, A. Mckee, J. R. Katzenbach.
*(17) Dr. Emilio Mira y López, MD: Guia de la salud mental, pag 18, Edit. Oberon.

Cont Section I: How To Manage "Yourself" (Soft Skills):

This is how our Mind works:
The thoughts of our mind*
are converted into words,
And those words will become our:
Behaviors, Habits and
Values = Ethical, Moral and Religious;
Which will define our personality (super-ego) and our
Success in life.

*Our Mind contains our :
Super-Ego = psychological or **Social environment**,
Where we: Grew, lived and developed in life;
and this is where we:
Learned, Copied and **Imitated**
Our examples or models of:
Behaviors, Habits and
Values: Ethical, moral, religious and cultural;
Which will define our Personality and our
Success in life.

Consequently, always remember that,
To Modify the personality we must:
ANALYZE, COMPARE & APPLY
Everything that we read in sections I & II of this book and:

BE VERY SINCERE
WITH OURSELVES.

This is the golden rule of Psychology,
that will help us To Modify the personality
{Modify the Super-ego and the Emotional Intelligence};
To achieve sustainable **success** at the:
Professional, Emotional and Financial Level.

This book contains practical management wisdom,
From brilliant world renowned: Presidents (CEOs),
Professionals (Contractors, Engineers, Architects, Inspectors & Unions),
Universities, and MDs mentioned;
To help us Improve **How we Manage "Ourselves."**

HAPPINESS IS RELATIVE !!!

Practical applications relating to Emotional Intelligence:

Comments by Harvard's (HBS) weekly newspaper by
Jehan de Fonseka (editor-in-chief of the Harbus):
"First test people for social & emotional intelligence;
Back to basics...How well do they care about each other."

"The idea behind this is that:
Good Leadership begins with self-knowledge.
The problem, with some leaders today,
Has little to do little to do,
With their ability to crunch numbers,
But rather a **lack of values;**" thus, it is advisable,
To modify the super-ego* and the Emotional Intelligence.

"When you think about the:
Biggest failures of corporate executives,
They are not necessarily technical failures,
But **ethical** ones;" urgently, it is advisable,
To modify the super-ego* and the Emotional Intelligence.

"If "some" of our business leaders:
Had more insight into their own:
Strengths and weaknesses,
It would have avoided the **excessive greed**
That led to the 2008 financial crisis;
Consequently, we need urgently more:
Training & practice in:
"Emotional & social intelligence"
{It is advisable, To Modify the super-ego* and the Emotional Intelligence}
To get back to basics....
To care more about each other."

Cont Section I: How To Manage "Yourself" (Soft Skills):

To achieve an adequate "maturity"
We must abandon the un-realistic objectives of:
Prepotence (arrogance) or illusions; otherwise,
To be able to avoid later delusions;
Therefore, it is advisable:
To avoid un-realistic dreams, and
Accept who we really are, and
Accept life for what it really is, without un-realistic illusions,
To avoid big delusions {setbacks or failures}; thus,
Saving that extra energy to achieve our:
Realistic, achievable dreams.

In conclusion: It is advisable, to observe and follow
the golden law of psychology:
Be truthful and sincere with yourself.

Remember what the:
world famous Dr. Erich Fromm, MD,
mentions in his book
"The Art of Listening" pags (68,69):
"Only a Fundamental transformation of our
Personality system,
Can produce a significant change of our **character**;
In other words:
We must **change not only in one aspect,
but in all our aspects of our Personality system;**
that is, the way we:
Think, act, feel, move .... and everything else,
because, **one isolated Emotional change,
Never produces a lasting effect."**
{It is advisable, To Modify the super-ego* and the Emotional Intelligence}

According to the world famous psychiatrist
Dr. Emilio Mira y Lopez, MD,
the **super-ego*** should be called **Anti-ego**, because,
There is a sector, of psychological forces hostile to the Ego."*(5)pag202

The **super-ego*** is the psychological environment, or
Social environment, where we:
Grew, lived, studied and **Imitated** the
Behaviors and Values: Ethical, moral, religious and cultural.

Type "A" Personality:

"Type "A" behavior pattern is an action-emotion
complex, that can be observed in any person, who is
Aggressively involved, in a chronic, incessant struggle:
"To achieve more and more in less and less time"
{Hurry sickness},
Very frequently, leads the type "A" to:
His early demise,
Afflicted from coronary heart disease."*(4)

"Persons possessing this pattern are also quite prone
to exhibit a free-floating:
Hostility."*(4)
{It is advisable, To Modify the super-ego* and the Emotional Intelligence}

"This bring us then, to the key reason for the
"Insecurity" of the Type "A" personality:
he has staked his innermost security, upon the
Pace of his **status enhancement.** This pace in turn
depends, upon achieving a maximal number of things,
in a minimum amount of time."*(4)
{It is advisable, To Modify the super-ego* and the Emotional Intelligence}

*(4) Treating Type "A" Behavior & your Heart,
by: Meyer Friedman, MD & Diane Ulmer, RN, MS.,
Published by Fawcett Crest, NY., Pags (84,87,93,229)

Cont Section I: How To Manage "Yourself" (Soft Skills):

Cont. Type "A" Personality

"Most type "A" possess so much aggressive drive
that it frequently evolves into a free-floating
Hostility."*(4)

"Perhaps, the prime index of the presence of
Aggression or hostility
in almost all type "A" persons,
is the tendency always
To compete with or to challenge **other people,**
whether, the activity consists of a:
sporting contest, a game of cards, or
a simple discussion."*(4)
{It is advisable, To Modify the super-ego* and the Emotional Intelligence}

"There are some persons, whom we consider
type "A", not because they are engaged in a
struggle, to achieve a maximal number of goals,
in a minimal amount of time, but because they
are so **hostile,** that they are almost continuously
Engaged in a struggle against another person."*(4)
{It is advisable, To Modify the super-ego* and the Emotional Intelligence}

"Type "A" individuals do tend to seek each other
out socially, despite the fact, that often their free
floating hostility and excessive competitiveness,
sometimes, converts their social meetings into
War meetings."*(4)
{Convenient To Modify the super-ego* and the Emotional Intelligence}

*(4)   Treating Type "A" Behavior & your Heart
by: Meyer Friedman, MD &Diane Ulmer, RN, MS.,
Published by Fawcett Crest, Pags.(95,227)

27

Also Dr. Bernard Jensen *(5) mentions, that
"To have a long lasting and enduring happiness,
We must combine and maintain an excellent
balance, of our four (4) main basic activities,
which are the following:

To obtain our
HAPPINESS
We must balance
Our four (4)
Basic activities:
}
      1) **Mental,**
      2) **Physical,**
      3) **spiritual and**
      4) **Material**

It is the proper balance of these four (4) activities
that makes for the happy wholesome life. "*(5)

To improve how to manage ourselves,
We need the **HUMANITIES,**
To understand ourselves better,
To modify all the deficiencies of our Super-Ego,
and achieve sustainable success at:
The Professional, Emotional and Financial Level.

Thus, "It is for this reason, that management will be
more and more, the way through which the:
HUMANITIES
Will again acquire worldwide renown,
To produce an impact on business, and will
Become an important field to learn & apply to
"Practical" business, according to worldwide
Management adviser P. Drucker."*(1)

*(5) You can Feel Wonderful & Enjoy it Now,
        By: Dr. Bernard Jensen, D.C. Pags.(48-49).
*(1) Management: Tasks, Responsibilities and Practices,
        By: Peter F. Drucker, Pag. 244

Consequently, in this book we will be using some of the knowledge from "some" of the following areas of:

HUMANITIES: {
Psychology,
Psychiatry,
Philosophy,
Religion,
History and
Management

From now on, **ANALYZE, COMPARE & APPLY**
Everything that you read and:

BE VERY SINCERE WITH YOURSELF.

This is the golden rule of psychology:
To be able, **To Analyze, Compare & Apply**
Sections I & II of this practical book, and
Apply the wisdom of the world renowned:
Presidents (CEOs), Professionals (Engineers & Inspectors),
Contractors, Universities, and MDs mentioned;
To achieve sustainable success at
The Professional, Emotional and Financial level
{It is advisable, To Modify the super-ego* and the Emotional Intelligence}

The "STRUCTURAL" COMPONENTS of
The "PERSONALITY":
1) The ID   2) The EGO   3) The SUPER-EGO

*The ID: { Instincts, Impulses, Passions } Irrationals & Un-Conscious

Controls the (pain & pleasure)

The EGO: {

Perception of: Reality & REASON

Tries To Regulate & Control the ID

}

SUPER-EGO:*

{Moral Censor

{ Habits, Examples, Beliefs, Teachings Behaviors & Values,* of Our: } { Parents, Relatives, Teachers, Neighborhood, Religious Advisers, Friends }

Psychological or **Social Environment,**
Where we: Grew, lived, studied and **Imitated**
The Behaviors & **Values:**
Ethical, moral, religious and cultural.

According to the world famous Psychiatrist
Dr. Emilio Mira y Lopez, MD:
The super-ego should be called Anti-ego, because,
There is a sector, of psychological forces hostile to the Ego;*(5)pag202
That needs to be modified urgently; to achieve success.

Super-ego = In psychoanalysis, is that part of the psyche which:
1) Is critical of the self or Ego, and
2) Enforces moral standards, at an unconscious level,
{**If** we have modified all the deficiencies of the Super-Ego};
Then, it blocks the irrational unacceptable impulses of the **Id.***

Summary: Hypothesis
How To Achieve a lasting Personality {Character} "Modification":
The Super-Ego = In psychoanalysis, is that part of the psyche which:
  1) Is critical of the self or Ego, and
  2) Enforces moral standards, at an unconscious level,
     {If we have modified all the deficiencies of the Super-Ego};
     Then, the Super-Ego blocks the irrational unacceptable impulses of the Id.*

*The **ID:** $\begin{cases} \text{Instincts} \\ \text{Impulses} \\ \text{Passions} \end{cases}$ Irrationals & Un-conscious

Controls the (Pain & Pleasure)

To enable, our super-ego to block the irrational:
Impulses, instincts or passions of the Id,*
It is advisable, to modify all the learned and imitated
Deficient behaviors and values of the super-ego pertaining to:
The Fear, Love, Duty and
Anger: Hatred, enviousness, prepotency, bad temper, and
Egotism** = Arrogant, or abusive or disrespectful behaviors.

Also it is advisable to modify -if present-:
The hyper-aggressiveness, hurry-sickness, Mal-intentioned humor;
and lastly -if present-,
The Un-conscious self-destruction of the:
Career, business, marriage or **life**
{Mentioned by Dr. M. Friedman, MD, and also by the world famous,
Dr. Emilio Mira y Lopez, MD; in his book: "The Four Giants of the Soul"}.

Thus, we must continue improving our **Inner** wealth & health,
To increase or retain our **Outer** wealth & health;
To be able, **"To improve or regain 40%** of our:
Performance, productivity, leadership and earnings of any
Corporation, contractor or profession."
{According to H. Geneen, ex-CEO of ITT}.*(2)

Thus, to achieve a lasting personality modification, it is advisable to apply,
the advice given in the book "The Art of Listening, Pags.(68,69),
by the world famous Dr. Erich Fromm, MD:
"Only a fundamental transformation of our personality system,
Can produce a significant change of our character; that is,
the way we: Think, Act, Feel, Move .... and everything else,
Because, **one isolated Emotional change,
Never produces a lasting effect"**
{It is advisable, To modify the super-ego* and the Emotional Intelligence}.

31

EGO DEFENSE Mechanisms, TO AVOID REALITY:

SELF-DECEIVING EGO DEFENSES Mechanisms, TO AVOID REALITY:

The child,
The adolescent &
The adult,
in order
To Avoid PAIN &
Obtain PLEASURE,
Utilize the following
DEFENSE
Mechanisms:

{

RE-PRESSION,
REACTION FORMATION,
PROJECTION,
SUBLIMATION,
INTELLECTUALIZATION,
IDENTIFICATION,
DENIAL,
SELF-DESTRUCTION,
INHIBITION and
RE-GRESSION.

REPRESSION :

It is the process of "excluding" & rejecting from consciousness
a thought or feeling that causes: pain, shame or guilt.
It denies all or part of the mental conflict, but by doing it,
It does not suppress the conflict, it only goes into
The unconscious, where it continues on acting "unconsciously".

REGRESSION :

Sometimes a frustrated individual "un-consciously"
seeks to return to an earlier, more secure period of
his life, such as a childish behavior, which he had
used before, to obtain his: childhood wishes.

REACTION FORMATION:

The development of opposite acceptable attitudes
that contradict his un-conscious wishes, for example
"puritanical" attitudes.

EGO DEFENSE Mechanisms, TO AVOID REALITY:

SELF-DECEIVING EGO DEFENSES Mechanisms, TO AVOID REALITY:

PROJECTION:

To blame "other" people -or even things-
for failures that are of "our own making".

SUBLIMATION:

In sublimation a need which
can **not** be satisfied directly,
then, the individual accepts some:
alternate goal, socially acceptable,
as an outlet of expresion of:
a sexual urge supresed,
or marital problems not resolved or without solution.

In sublimation, some of the alternate goals, socially acceptable
utilized by individual are the following:
artistic activities, all forms of;
work-science and business,
"Sometimes" derived from the sublimation of sexual energy.

The problem in the work-business environment is that, sometimes,
there is so much accumulated "anger"
{due to marital problems not resolved or without solution},
that the sublimated activity selected,
might lead to disastrous decisions,
regarding the future of the:
Country, corporation, contractor, project or profession.
{It is advisable, to modify the super-ego* and the Emotional Intelligence}

EGO DEFENSE Mechanisms, TO AVOID REALITY:

SELF-DECEIVING EGO DEFENSES Mechanisms,
TO AVOID REALITY:

INTELLECTUALIZATION:

Another way, to compromise with problems,
Is to intellectualize them, thus partially divesting them of
Personal significance or painful feeling.

There are three (3) basic mechanisms of Intellectualization:

1) Rationalization or Excuse Making

2) Isolation or the use of logic tight compartments

3) Undoing or ritualistic "cleansing" Behavior.

DENIAL:

It assumes that a feeling or thought
Is **not** Dis-Agreeable, when in reality
It is Dis-agreeable.

INHIBITION:

The individual stops doing or acting,
It is the suppression or restrain of behavior.

EGO DEFENSE Mechanisms:
SELF-DECEIVING EGO DEFENSES Mechanisms,
TO AVOID REALITY:

IDENTIFICATION:

In this mechanism, the frustrated individual
incorporates into his own personality structure,
the achievements or qualities of those who
frustrate or threaten him.
from this mechanism
comes the popular saying:
"If you can **not** beat them, join them."

The previous information is very important
To Learn:

HOW TO MODIFY
THE PERSONALITY,
OR THE SUPER-EGO

It is advisable, to modify the super-ego* and the Emotional Intelligence,
To be able, To Analyze, Compare & Apply
sections I & II of this practical book, and
Apply the wisdom of the world renowned:
Presidents (CEOs), Professionals (Engineers & Inspectors),
Contractors, Universities, and MDs mentioned;
To achieve sustainable success at
The Professional, Emotional and Financial level.

DON ZEUS: The SILENT CO-AUTHOR

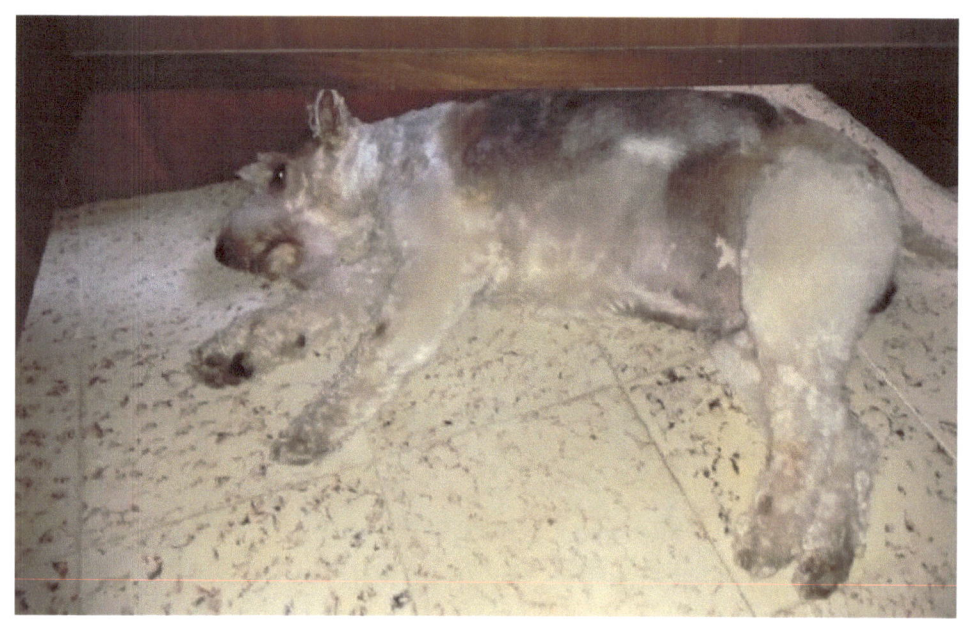

Definition of "Wealth"

Wealth is composed of two (2) elements:
1) **Inner** = Happiness = Satisfaction with life
   Modified Super-Ego*and Emotional Intelligence
   Excellent Personality (Character)
   Inner peace, harmony = **60%**

2) **Outer** = Assets, Positions,
   Titles, Diplomas & money($$$) = <u>**40%**</u>
                                    100%

⎫ **Wealth**

Summary: Improving our **Inner** wealth,
Increases or preserves our **Outer** wealth.

Wealth is composed of two (2) elements:
1) The **Inner Wealth = 60%** of our total wealth,
   It gives us Happiness = Satisfaction with life, inner peace,
   to be able, to apply the golden rule of psychology:
   "Be very sincere with yourself,"
   To modify the Super-Ego,*and the Emotional Intelligence,
   To obtain an excellent personality (Character), to be able:
   a) To avoid mistakes, accidents, or losses, and
   b) Improve our **Inner** wealth; thus,
      Increasing or preserving our **Outer** wealth, and also our:
      Mental, physical and financial Health.

2) The **Outer Wealth = 40%** of our total wealth,
   Which is always exposed to the mortal trap, mentioned
   By the ex-President of ITT, H. Geneen: sometimes,
   "More fortunes (or careers) **are destroyed by success than by failure,**
   Frequently, generating Egotism"**(2)
   = Arrogant, abusive or disrespectful behavior; sometimes,
   Causing harm to others or themselves.

Conclusion: If we utilize our improved **Inner** wealth,
To modify the Super-Ego* and the Emotional Intelligence; then,
we will be able, to increase or preserve our **outer** wealth, and also
Succeed at the emotional, professional and financial level.

According to the world famous, Dr. Emilio Mira y Lopez, MD:
*The super-ego = Anti-ego, because there is a sector, of psychological forces
Hostile to the ego. The super-ego* is the psychological or **Social environment,**
where we: Grew, lived, studied and **Imitated** the Behaviors and
Values: ethical, moral, religious and cultural.

**(2) Managing, by Harold Geneen (Ex-CEO ITT), Pag (184-5) Pub. by Avon Books.

37

Cont Section I: How To Manage "Yourself" (Soft Skills):
Matching points (*=*) of the **Management styles** of:
*Segundo Cristancho J. & Company {during 67 years}, and
*"Mr. Harold Geneen, Ex-CEO of ITT
{Helped a **$766** million telephone company grow into a
$22 Billion multinational conglomerate, according to
His book: Managing, Pub. by Avon Books}:*(2)
Several executives, that worked with Mr. Geneen,
graduated to successful careers as Fortune 500 CEOs:

*=*1) **Plan, direct and control:** "Know every facet of the businesses" and
= 2) Treat people with respect & admiration and listen with empathy,
have an open door, but avoid incompetence and neglect.
*=*3) Urgently, If present, modify the egotism,**
To be able, "To improve or regain 40% of the:
Quality, safety, performance, productivity,
Innovation, competitiveness and earnings of any
Corporation, contractor or profession
{Mr. H. Geneen, Ex-CEO of ITT}."*(2)".
{Modify the super-ego* and the Emotional Intelligence,
applying sections I & II of our practical book}.
*=*4) "Update the plans, associates must update them too.
Use what worked, but changed when necessary."
*=*5) "Be a: {Hands on, present, active and a responsible} manager."
*=*6) "Know:{your business, what works, your people, who works}
*=*7) "Manage with respect & admiration (w/o egotism)."**
"Performance and results must be achieved,
If you don't achieve those results, you're not managing"
{To Improve your management, apply sections I & II of our book}.
*=*8) "Put in the time and dedication to work things out."
Never delegate the responsibilty of controlling a
Task or a goal, to be done, with excellent:
Quality, safety, on time and under budget.
*=*9) "When a client reports a complaint or suggestion,
Communicate it to management ASAP, and fix it asap,
according to the plans, contract and specifications,
to comply with the client wishes and desires; thus
always having a satisfied and loyal client."

*Also, applied at: **Segundo Cristancho J. & Co.** during 67 years.
*(2)Managing: Harold Geneen, ex- CEO, ITT, Pub. by Avon Books.
**Egotism = Arrogant, or abusive or disrespectful Behavior

The fight against BAD TEMPER:

It is advisable, to Modify the super-ego* and the Emotional Intelligence,
According to world famous, Dr. Emilio Mira y López, MD,
in his excellent book "The four big Giants of the Soul",
"Fear, Anger, Love and Duty"*(5), he mentions the following:

"It is convenient to remember that, anger comes from fear,
and anger blinds the sight and the understanding; thus,
If we want to avoid being a victim of that angry impulse,
Then we have to start"*(5)
—— "**Knowing ourselves better** —— because the more
We know about ourselves, the easier it will be to identify:
1) The good natural talents to be developed, and
2) The personality's deficiencies that will be convenient to eliminate,
To be able to modify the Super-ego*and the Emotional Intelligence;"*(19)
"Thus, the afflicted person can stop suffering, and
That person will stop making other people suffer."*(5)

"The bad temper is a sign of deep anger,
Showing signs of insecurity, lack of self control, and
lack of faith in his own capabilities."*(5) pags.124,5

"Be careful, the bad temper is contagious,
A single person with bad temper can infect the whole group;"*(5)
The brilliant ex-CEO of G. E., Jack Welch mentions that:
"A single person with a bad temper
Can destroy the work of a whole department, and
Can infect the whole corporation or profession."*(19)

Summary: "It is advisable, to know ourselves better,
To modify the Super-ego* and the Emotional Intelligence,
To diminish or avoid the internal anger; to be able,
To live in peace with ourselves and others."*(5)

*(5)Dr. Emilio Mira y López, MD: "The four big Giants of the Soul",
"Fear, Anger, Love and Duty", pags.124,5, Publ. Lidiun.
*(19) The Real MBA, by Jack & Suzy Welch, pub. by Harper, N.Y., N.Y.

According to the ex-CEO of G.E. Jack & Suzy Welch,
They mention the following in theirs book
The Real Life MBA*(19):

1) "Real **Leadership** is based on **2 Ts: Truth & Trust.**
The **Leader** also utilizes the 4 E: Energy, Energize, Execute & Edge;
To have the courage & passion to make tough and risky decisions,"*(19)
Applying an excellent modified Super-ego* and Emotional Intelligence.

"A Leader also has excellent **discernment,**
To understand, appreciate and respect,
To analyze situations, problems or decisions"*(19)
{Excellent Modified Emotional Intelligence and super-ego*}.

"Table: Typical Evaluation of Employee: Performance and Behavior:"*(19)

EXCELLENT / GOOD	NEEDS IMPROVEMENT
A) 20% = Are super stars	C) 10% = Low performance
B) 70% = Are average	*(19)

*(19)The Real life MBA, by Jack & Suzy Welch, pags 67,125,151, Pub. Harper, NY.,NY.

Cont. Section I: How To Manage "Yourself" (Soft Skills):

Cont. According to the ex-CEO of G.E. Jack & Suzy Welch,
They mention the following in theirs book
The Real Life MBA*(19):

2) **"The Innovation** = Is the integral sum of small innovations = Σ = $\square$.
The Innovation = Σ = $\square_{n=1}^{n=\text{infinite}}$ = Innovations =Σ from n=1, to n=infinite."
"Daily Innovation = Everybody looking for better ways to do things;
To achieve the best product or service, with
The best: quality, safety, and price."*(19)

3) **"Strategic planning:** Flexible and agile, because the market conditions
Move and change all the time: Locally, nationally and globally.
To apply the best practices and obtain a competitive advantage,
It is advisable, to select the best personnel, with the best theoretical
And practical knowledge, and assign them to the most di□cult jobs;
To be able to optimize the Innovation and competitiveness:
Locally, nationally and globally."*(19)

4) **"Team work: Always allow every member,
To express theirs solutions or ideas,
To improve the innovation and competitiveness;**
This strategy will help, to give dignity to the team members; thus,
They will feel part of the team, corporation, contractor or profession."*(19)

5) **"** The brilliant ex-CEO of G. E., Jack Welch mentions that:
"A single person with a bad temper
Can destroy the work of a whole department, and
Can infect the whole corporation, contractor or profession."*(19)
{It is advisable, to Modify the Emotional Intelligence and the super-ego*}

Summary: "It is advisable, to create a pleasant working environment,
To make everybody feel appreciated and recognized,
For theirs performance and behavior; thus,
Improving productivity and personal satisfaction."*(19)

*(19)The Real life MBA, by Jack & Suzy Welch, pags 67,125,151, Pub. Harper, NY.,NY.

Narcissism :

According to the world famous psychiatrist
Dr. E. Fromm, MD:
"A narcissist is a person who thinks that
"only" what he:
Thinks, has, is or does is important,
and what the rest think, have, are or do is not."

The psychologist F. Dorsch defines the narcissist
as "a person who is in love with him/herself
{regarding what he thinks, has, is or does},"
which is sometimes due to low self-esteem or
psychological in-maturity.

The psychiatrist, Dr. E. Fromm; MD,
gives a good definition of a:
"Narcissist" { OVER-values him/herself &
As one who: { UNDER-values others."

We will mention the inter-personal friction
Narcissists cause themselves and others,
such as: **not** being liked by many people, because
they are perceived as being:
ARROGANT,
PREPOTENT   Money,
Regarding   Intelligence or
Their own: Position
Sometimes causing people to reject them or
wish them the worst luck in their professions or
businesses, or personal relations. Thus, there is an
urgent need, to modify this excessive narcissism; thus,
It is advisable, to modify the super-ego* and the Emotional Intelligence,
To avoid the "grave consequences" in theirs:
Careers, businesses, or personal lives !!!.

Awaken=Awareness*(9)

Awaken= Awareness:

"A term used in
Psychology to describe the
Person who has been able
To modify his/hers:
    "Narcissism";
Thus, eliminating the illusion of
Having an indestructible ego and
Diminishing his/hers suffering.

Only by awakening from those

Illusions
Is one able to know:

The Reality
of his/hers own:

Sickness,
Old age,
Death and the
Impossibility of
Achieving the
Unreal Illusions
Desired."*(9)

*(9) E. FROMM: The HEART of MAN, Pag. 101
Pub. by: Harper & Row. El Corazón del Hombre:
Pags.(68-133)

LIFE's SECRETS

The "Secrets" for Obtaining and **Retaining :** { FINANCIAL, PROFESSIONAL and EMOTIONAL } Sustainable **WEALTH**

"Lifes' Temptations: { To be somebody To have: } Power and Money*(11)

"This quest make people suffer, making life a continuous search toward God; that is why, Human beings take their sufferings to church, where they find strength, to overcome and continue life's struggles with greater strength and hope."*(11)
To be able to succeed in life, we must always:
"Have GOD as a permanent partner."*(11)

"To Achieve Sustainable **Wealth:** { MATERIAL, SPIRITUAL, PROFESSIONAL, EMOTIONALand FINANCIAL } Obtained by the One who Listens, Understands and Applies the Advice of God"*(12)

*(11) Pag 732, Comentarios B. LatinoAmericanos.
A.T II; Editorial V.D.; España.
*(12) Pag 343, Comentarios B. LatinoAmericanos.
N.T; Editorial V.D.; España.

Depression:

The famous psychologist Dr. A. de Mello* recommends
The following, to avoid or diminish depression:
"Never identify yourself with that feeling; i.e.
Never identify yourself with the depression;
Never say: I am depressed, instead
you should tell yourself:
the depression is here,
There are some hurt feelings here right now,
but they will pass, just leave them alone,
It will pass, everything will pass, everything."!!!
{Always seek the best medical assistance in **all** cases regarding your total health}

"Your depressions and emotions
Do not have anything to do with your happiness,
That is your illusion;
Because you are defining yourself in terms of
that illusive feeling,
Which does not correspond to reality;
Because it is an illusion, and it is a mistake."
{Always seek the best medical assistance
in ALL cases regarding your total health}

"If you feel emotions, get ready for depression.
Never identify yourself with the depression;
It does not have anything to do with yourself;
just tell yourself: there is a depression here,
but it will pass, everything passes, everything.
Never define your essential being in terms of
a depression or loneliness."

{Always seek the best medical assistance in ALL cases regarding your total health}
Warning: This book is not intended to take the place of medical advice
from a trained medical professional; thus, always seek the best medical
Assistance in ALL cases regarding your total health. Neither the authors,
Nor the publisher of this book, nor the publishers of the books mentioned
in this book take any responsibility if this warning is not followed completely.

Guilt :

The famous psychologist Dr. A. de Mello advises the
following to diminish or to abolish "previous" **Guilt,**
After having acknowledge remorse,
Tell yourself that,
You are no longer those previous guilty feelings,
because you were ill-programmed or
Hypnotized,
You were not preprogrammed adequately; therefore,
you were not yourself, because you were hypnotized:
Acting under the influence of erroneous guidance,
Repeating those old bad habits, learned from your:
Previous un-modified programming.

In other words:
After having acknowledge your remorse, tell yourself:
You were not responsible, you can not blame yourself
for your previous bad actions, because
you were not preprogrammed adequately; therefore,
you were not yourself, because you were hypnotized:
Acting under the influence of erroneous guidance,
and you should "pardon yourself immediately"
To be able, to Re-program yourself.

After having acknowledge your remorse, you should repeat to yourself:
"You were not to blame for your previous bad actions,"
so that you can awaken, from your state of deep-seeded hypnosis;
Allowing you, **to re-program yourself, and live a better life.**
{Always seek the best medical assistance in ALL cases regarding your total health}

Warning: This book is not intended to take the place of medical advice
from a trained medical professional; thus, always seek the best medical
Assistance in ALL cases regarding your total health. Neither the authors,
Nor the publisher of this book, nor the publishers of the books mentioned
in this book take any responsibility if this warning is not followed completely.

Look HOW ANIMALS RELAX,
"Sometimes" BETTER than HUMAN BEINGS !!!

Cont Section I: How To Manage "Yourself" (Soft Skills):

The book: *Snakes in Suits: When Psychopaths Go to Work by
 Drs. R. D. Hare and P. Babiak, Phds.*(7), mentions that
 Some of the red flags of the psychopaths might be:

1) Superficially charming,
2) Pre-potent, and/or narcissists,
3) Lie and manipulate people easily,
4) Lack remorse & empathy
5) Are cold, in-considerate and
          Mean, and do not accept responsibility.
6) Only care about: Money & Position,
7) Are irritables, impatients & impulsives: (the 3 I)
8) Are Bullies/or abusive with their subordinates,
          But very nice to their superiors,
          Thus, creating: Low morale or
          Loss of: (productivity or satisfaction or)
                    Market share & Earnings,
          and also high personnel turnover.*(7)

"Dr. S. Freud tried unsuccessfully to explain."*(8)
          Why the psychopath has not yet resolved,
A very important unconscious childhood's issues
{Relating to the unresolved, unconscious hostility toward one of the parents}.

According to Dr. S. Freud,
The child unconsciously hates one of the parents;
Thus, lacks that important parental censorship activity, and
Continues braking all the norms or duties taught,
          Lacking - according to Dr. S. Freud-
          The so called Super-Ego* {or Moral Censor}"
{Mentioned by the world famous psychiatrist,
          Dr. Emilio Mira y Lopez, MD, in his book,
          "The four Giants of the Soul"
"The Fear, Anger, Love and Duty", pag. 201-2, publ. by Lidiun}.

"Sometimes the psychopath has brain's abnormalities in the sections
of the Ventrolateral, orbitofrontal Cortex and Amygdala."*(8)

*(8)The Brain and Emotional Intelligence: New Insights, By Harvard's Psychologist,
Dr. Daniel Goleman,Published by More than Sound, 2011. I recommend this Important digital book.

Cont Section I: How To Manage "Yourself" (Soft Skills):
The advisable School and University "formation":
The world famous psychiatrist:
Dr. Emilio Mira y López; MD,
Mentions in his book "Guide to Mental Health" that:
To achieve a modified personality, it is advisable for
"The school or University, to place more emphasis on
The "formation" than on the "instruction" of its students,
Teaching them, from the early years,
How to solve, with calmness and decision,
The difficult situations,
They will encounter in real life, and in theirs personal lives."

"Dr. Emilio Mira y Lopez, MD, suggests to follow two (2) steps,
To achieve the proper educational "formation":
1) Create in each student a feeling of
Inner peace and trust in oneself.
2) Avoid conflict between
Real ambitions and realizations,
Regarding wishes and successes,
In other words, achieving what was intended originally,
To avoid falling prey to "imaginary achievements";
That may result later in addictions to:
Gambling, drinking, imaginary fictions, or neurosis."

"To achieve this mission, it is necessary,
The collaboration between school and family;
and the teachers, must work together
with the students' family to convince them that:
A person is worth not for what he knows and feels,
But for what "good" he is capable to do,
With what he knows and feels."

"The true educational role of the school or university is
To make people understand, that the real worthiness of a person is
To be the owner of his own destiny, and to know:
How to procure first, the wellbeing to everybody, before his own;
that is, the real educational formation of the school or University,
To collaborate in the fight against the present social diseases and vices."

Cont Section I: How To Manage "Yourself" (Soft Skills):

We will now describe some commentaries, very
Adequate at this moment, from the book called:

The Four (4)
SOUL'S Giants:
{
Fear,
Love,
Duty and
Anger
}

This is a book written by the world famous
Dr. Emilio Mira y López; MD, and
It mentions the following advice, regarding
The person who is
Pre-potent or Arrogant.
This person is full of fear, trying to show
the opposite, trying to convince himself, that
there is no motive to feel insecure; because,
According to him he is more worthy than anybody else.

But if he has to keep repeating it to himself constantly,
It is because, deep down, he not only doubts it, but
He is convinced that he is not worth too much;
Consequently, he tries to appear as a:
Pre-potent or arrogant person,
To cover up, his real opposite personality,
with very low self-esteem and full of fear !!;
{Fragile Super-Ego,* It is advisable,
To modify all the deficiencies of the Super-Ego*
Including the egotism**}.

According to the world famous,
Dr. Emilio Mira y Lopez, MD:
"The super-ego should be called Anti-ego, because,
There is a sector, of psychological forces hostile to the Ego."*(5)pag 202

The super-ego also defines the psychological or **Social environment,**
Where we: Grew, lived, studied and **Imitated** the Behaviors and
Values: Ethical, moral, religious and cultural.

**Egotism = Arrogant, or abusive or disrespectful Behaviors.

Cont Section I: How To Manage "Yourself" (Soft Skills):

HOW to Improve **your professional SALES or SERVICES:**
To implement the following sales recommendations,
It is advisable first, to modify the Super-Ego* = Anti-Ego;* such as:
Anger, or prepotency, or enviousness, or bad temper or egotism.**

The following are the recommended keys,
To obtain or retain clients and
Improve your professional Sales or Services:
1) **Enthusiasm,** to obtain or retain prospective clients:
Always, visit them personally, and keep in touch regularly
Showing the client how your products or services can be
The best solution to the client's needs and wants.
2) **Listen** with improved:
Empathy, respect, admiration and Emotional Intelligence;
To be able, to learn what are the client's needs and wants,
and always, looking out, for the best interest of the client.
3) Always have an excellent product or professional service,
Offering the best: quality and safety,
At the best price possible
{Without sacrificing safety or quality},
4) Excellent Knowledge of your:
a) Product or Profession and
b) Professional Association or industry as a whole, and
5) Always act with the best personal
Honesty, integrity and dependability;
After all, people buy
Services or **products** from people

According to the world famous,
Dr. Emilio Mira y Lopez, MD:
*"The super-ego should be called Anti-ego, because sometimes,
There is a sector of psychological forces hostile to the Ego."*(5)pag202

*The super-ego also defines the Social or psychological environment
Where we: Grew, lived, studied and **Imitated** the
Behaviors and values: ethical, moral, religious and cultural.

**Egotism = Arrogant, or abusive or disrespectful Behaviors.

Cont Section I: How To Manage "Yourself" (Soft Skills):
According to the world famous psychiatrist:
Dr. Emilio Mira y Lopez, MD, mentions the:
Definition of ENVY:

"Envy carries in its root, a considerable charge of
Anger;
Thus, you must
remember this, when you notice,
that somebody feels envious of yourself,
showing you, that he suffers an
intense internal **anger,**
sometimes due to his low self-esteem."

The Psychologist: Dr. Friedrich Dorsch mentions
the following definition of
Envy:
He defines "Envy as a feeling of dis-pleasure
due to the happiness of somebody else,
because the envious person is:
unable to share the happiness
of the one who is successful."

Therefore, there is a better way to help everybody,
It is advisable, to modify your personality
{Modify the super-ego*and the Emotional Intelligence};**
To improve your Emotional Intelligence,
so that your success will not be impaired at:
The professional, emotional and financial level.

**"Be very sincere with "yourself"
To modify the super-ego* and the Emotinal Intelligence;
To be able, **To ANALYZE, COMPARE & APPLY**
the advice of the world renown professionals,
Mentioned in sections I & II of this practical book.

THE MAL-INTENTIONED "Humorist"

The psychiatrist: Dr. Emilio Mira y Lopez; MD,
Describes the bad or mal-intentioned humorist,
as an angry failure
who has a deep **fear,** and
"Sometimes" resorts to bad jokes or humor,
To be able to say it humorously,
what he is not able to say it seriously.

In this case, the bad jokes or fun made at
the expense of somebody else, is made by a person
who is resented, who has a great deal of inner **anger;**
showing how much he despises himself, and "sometimes"
that person feels the same as the old court joker.

To protect yourself psychologically from the
Mal-intentioned humorist,
It is advisable, to start urgently, the process
To modify the Super-Ego* and the Emotional Intelligence;**
To prevent the mal-intentioned humorist
from impairing your success at the:
Professional, Emotional and Financial Level.

** "Be Very Sincere with "Yourself"
To be able, **To ANALYZE, COMPARE & APPLY**
The advice of the world renown professionals,
Mentioned in sections I & II of this practical book;
To be able,
To modify the super-ego*and the Emotional Intelligence.

Cont Section I: How To Manage "Yourself" (Soft Skills):

Remember what the:
World famous **Dr. Erich Fromm, MD,**
Mentions in his book
"The Art of Listening" pags. (68,69):
"Only a fundamental transformation of our
Personality system
can produce a significant change of our **character;**
In other words:
We must change **not** only in one aspect,
But in **all** our aspects of our personality system
That is, the way we:
Think, act, feel, move .... and everything else,
Because, **one isolated emotional change,**
Never produces a lasting effect"
{It is advisable, To Modify the super-ego* and the Emotional Intelligence}.

Consequently, "It is imperative,
To acquire more insight into our own:
Strengths and Weaknesses,
To be able to modify our personalities
{Modify the super-ego* and the Emotional Intelligence};**
To achieve success at:
The Professional, Emotional and Financial level.

**"Be Very Sincere with "Yourself"
To be able,
To **ANALYZE, COMPARE & APPLY**
The advice given by the world renown professionals,
Mentioned in sections I & II of this practical book;
To Modify the super-ego* and the Emotional Intelligence.

TREACHEROUS PEOPLE
SOONER or LATER WILL BE DISCOVERED !!!!

La traición
Tarde o tempano va a salir a la luz.

TIME Management:

*"Part of the key to time management
is carving out time:
To think,
Strategically, pro-actively, longer-term,
as opposed to constantly reacting." *WSJ Feb 13/2012
(extinguishing fires)

*"Jeff Weiner, the CEO of Professional Network
Linkedin tells The Wall Street Journal
How he carves out time:
To think,
"Instead of just reacting to challenges."
"Spend a good percentage of your time with:
Customers and clients,
To keep you externally focused."*WSJ Feb 13/2012

HOW **TO DELEGATE** a task or a goal:
Always put in the time and dedication to work things out.
Never delegate the "permanent" responsibility of controlling
How a task or goal is achieved with excellent:
Quality, safety, on time and under budget;
that "permanent" responsibility of controlling everything,
Always pertains to all the contractors, professionals and the team.
{It is advisable, To Modify the super-ego* and the Emotional Intelligence}.

Cont. **TIME** Management:

"Microsoft's ex-CEO Steve Ballmer, creates a spreadsheet
To budget time for the year, allocating time for:
Meetings, travel and exploring new ideas.
"I've got a spreadsheet, it's got a budget—
my time for the year...
I give the budget allocation to my:
Administrative assistants,
They lay it all out and then
Anybody who asks for time, they say, Steve,
This is in budget, or it's not in budget,
How do you want us to handle it?."

What do chief executives (CEOs) do all day?
They spend about a third (1/3 = 33%)
of their work time in meetings.
As world famous management adviser:
Peter Drucker wrote:
"Meetings are by definition
A concession to a deficient organization.
for one either meets or one works,
One can not do both at the same time."*WSJ. Feb 13/2012

HOW **TO DELEGATE** a task or a goal:
Always put in the time and dedication to work things out.
Never delegate the "permanent" responsibility of controlling
How a task or goal is achieved with excellent:
Quality, safety, on time and under budget;
That "permanent" responsibility of controlling everything,
Always pertain to all the contractors, professionals and the team.
{It is advisable, To Modify the super-ego* and the Emotional Intelligence}.

DELEGATING a TASK or a GOAL:
"The great difficulty, the **type "A" personality** experiences,
Is delegating work to others,
This is due mainly to his belief, that all his activities:
Require his right hand and only his right hand.
Nothing can be delegated to anyone else,
This chronic refusal, grows out of the basic
"Insecurity"
of the type "A" subject*(4)
{Advisable To modify the super-ego* and Emotional Intelligence}.

"This brings us, then, to the key reason for the
Insecurity of the type A person:
He has staked his **innermost security** upon
the pace of his status enhancement.
this pace in turn depends upon;
A maximal number of achievements
Accomplished in a minimal amount of time;
inflicting upon himself an un-necessary amount of
high stress, which sometimes produces many
Cardio-coronary illnesses, and also,
Producing an un-necessary overloading to the
Central nervous system (CNS)."*(4)

HOW **TO DELEGATE** a task or a goal:
Always put in the time and dedication to work things out.
Never delegate the "permanent" responsibility of controlling
How a task or goal is achieved with excellent:
Quality, safety, on time and under budget;
That "permanent"responsibility of controlling everything,
Always pertain to all the contractors, professionals and the team.
{It is advisable, To Modify the super-ego* and the Emotional Intelligence}.

*(4) Treating Type "A" Behavior & Your Heart by:
Meyer Friedman, MD & Diane Ulmer; RN, MS.,
Published by Fawcett Crest, NY., Pags {241,2,4}

Cont Section I: How To Manage "Yourself" (Soft Skills):
This definition applies to all the professions pertaining to:
Engineering & Construction:
Contractors, Engineers, Architects, Inspectors and Unions.

Definition of a PROFESSIONAL:

1) A true professional possesses two types of knowledge:

Types of knowledge
{
1) The Professional or practical wisdom, and
2) The **one self's knowledge** =
Modified: { Emotional Intelligence and Super-Ego;*

To be able, To avoid mistakes, accidents or losses.

2)   Always applies the golden rule of psychology:
"Be very sincere with yourself,"
To modify the Emotional Intelligence and the Super-Ego* pertaining to:
"The Fear, Love, Duty and
Anger: Hatred, enviousness, prepotency, bad temper, and egotism,**
Mentioned by the world famous Dr. Emilio Mira y Lopez, MD
{In his book, "The four Giants of the Soul", Edit. Lidiun}."*(5)

3) Optimizes the relationships with others, to listen to them with
Empathy, respect and admiration; to implement the
Innovations and changes required,
To improve or recuperate 40% of the
Quality, safety, performance, leadership,
Innovation, competitiveness and earnings of the:
Country, corporation, contractor, project or profession.

4) A Professional continuously improves the Inner wealth;
To be able, to improve or preserve the Outer wealth.

On Nov. 1/202, Business Week mentioned that,
"Some Professionals are:
Hired for their excellent professional knowledge and
Fired for their Personalities" = Un-modified Super-Ego*and Emotional Intelligence.

According to the world famous,
Dr. Emilio Mira y Lopez, MD, the
*Super-ego = Anti-ego, because there is a sector, of
Psychological forces hostile to the Ego.*(5)pag 202

The super-ego is also, the psychological or **Social environment,**
Where we: Grew, lived, studied and **Imitated** the behaviors and
Values: ethical, moral, religious and cultural.
**Egotism = Arrogant, or abusive or disrespectful behaviors.
*(2) Managing, by H. Geneen, Ex-CEO ITT, Pag (184-5) Pub. by Avon Books.

Cont Section I: How To Manage "Yourself" (Soft Skills):

To Modify or
Re-Program
The Personality,
(Super-Ego)*
We have   Pre-judgement
To observe   or
Ourselves,   Desire To reform
And study our
Own reactions with
Others & things
Around us, Without:

In this way,
You will not feel forced to modify yourself,
and you will feel an internal peace or calm,
because, you will be gaining a profound
Self-knowledge, and you will feel:
More at ease with yourself,
Feeling an inner peace or harmony, and
very soon, you will begin to modify your:
Behaviors, habits and
Values: Ethical, moral and religious
= "Super-Ego",*
And you will become:
Happier, healthier and more productive,
To achieve sustainable success in life.

According to the world famous,
Dr. Emilio Mira y Lopez, MD:
*"The **super-ego*** should be called Anti-ego, because,
There is a sector, of psychological forces hostile to the Ego."*(202)
*The **super-ego*** is the psychological or **Social environment,**
where we: Grew, lived, studied and **Imitated** the
Behaviors and values: Ethical, moral, religious and cultural.

Cont Section I: How To Manage "Yourself" (Soft Skills):

The multi-billionaire oilman
John D. Rockefeller
Used to teach his fellow executives the following verse:

"An old wise owl sat on an oak tree;
the more he saw, The less he spoke;
the less he spoke, the more he heard;
Why can't we all be like that old wise owl ??."*(6)

TO MODIFY or
RE-PROGRAM
Ourselves,
We must have
A Mind:

Un-hurried, Calm, Serene,
Passively alert,
Without doing or thinking
More than one thing at a time,
To be able to:
Concentrate & Analyze
Each problem Individually,
Without distraction.

*(6) The Seven Sisters, by: Anthony Sampson,
Pag. 27., Edit.: Bantam Books, New York.

Cont Section I: How To Manage "Yourself" (Soft Skills):

I will mention some "practical" advice to apply
in your daily life, to avoid friction in your
Inter-personal relations:

Whenever there is an argument, generally due to minor,
un-important matters, always remember, the excellent
Advice from the Asian's wise men, who believed,
That whenever two persons are arguing that:

Each one of   { **RIGHT** }
Them was a   {   &   }
Little bit :   { **WRONG**   }

In other words, when the Asians are resolving
a judgment between two persons, A and B; then,
Most of the time, they will apply this rule:
Person A is right, but
Person B is not **wrong either !!!.**
This advice is given, by the famous Chinese
philosopher Lin Yutang, in his book:
My Country and my People, pag. 104.

In Conclusion:
It is advisable, to stop arguing for "minor"
un-important things.
If you modify the super-ego* and the Emotional Intelligence,
Then, you will always remember that, "sometimes",
The other person might be right,
and this technique will contribute to obtain more:
Friends, clients or contracts; thus,
It is advisable, to stop arguing & follow this excellent advice.

Cont Section I: How To Manage "Yourself" (Soft Skills):

At this time, It will be very convenient,
To remember some phrases, from the world famous
Psychologist and philosopher:
Dr. Anthony de Melo:
Concentrate on looking,
and some day you will be able to see!!!

Dr. A. de Mello also mentions:
Nothing has change,
Except: My attitude; } EVERYTHING
that is why: } HAS CHANGE

Remember what the:
world famous psychiatrist Dr. Erich Fromm, MD
mentions in his book
"The Art of Listening" Pags. (68,69):
"Only a Fundamental transformation of our
Personality system
Can produce a significant change of our **Character.**

In other words:
We must change not only in one aspect,
but in all our aspects of our personality system;
that is, the way we:
Think, act, feel, move .... and everything else,
because, **one isolated emotional change,**
Never produces a lasting effect."
{Advisable To Modify the super-ego* and the Emotional Intelligence}.

Warning: This book is not intended to take the place of medical advice from a trained medical professional; thus, always seek the best medical assistance in ALL cases regarding your total health. Neither the authors, nor the publisher of this book, nor the publishers of the books mentioned in this book take any responsibility if this warning is not followed completely.

Cont Section I: How To Manage "Yourself" (Soft Skills):

The Mind $\Big\{$ Interested $\Big\}$ Understand
Is calm or   Truth that,
Un-hurried   &   Modifies
When it is: $\Big\{$ Desires to: $\Big\}$ The Personality (super-ego*)

To Understand
   & Find your $\Big\}$ Silence & Observation,
     Real self,   Passively perceptive & alert,
   We have   with a: Flexible Super-Ego
To remain in:

The Modification or Re-Programming of our mind
Must take place in:

Our: Wishes,
Super-Ego   Desires,
= Moral Censor   Beliefs   Ethical,   Of our
= **Social Environment**   Values,*   Moral,   MIND
       Habits   Religious,
            Cultural
      *Values: Ethical, moral and religious

And **not** in: $\Big\{$ Have ($$$)
What we   or
      Wear

Cont Section I: How To Manage "Yourself" (Soft Skills):

We will mention a brief story, that will help us
To Become   Humble, and
the story is the following:
It happens at "Oxford" University,
A "newly" graduate told an old professor:
thank you very much professor:
with the education that I have received,
I feel that I am completely prepared for life !!!;
and the old professor answered him very humbly:
In my case my son:
Only until now I feel completely prepared for life.

Perhaps, this is a good time to mention, that even
If one has two (2) **University degrees,** the maximum
That they can teach us at the University is around
Twenty (20%) percent.
The other 80% must be obtained throughout:
Life's practical experiences, attending:
Seminars (locals, nationals, & internationals),
consulting with your colleagues,
Studying (keeping up-to date),
attending Post-Graduate School,
and doing your own R & D.*

It will be quite appropriate,
to mention the words of wisdom of
the world famous cardiologists:
Drs. Friedman, Rosenman, MDs, Ulmer & Associates:
**"We modern physicians and nurses
still have much to learn, and
much to be modest about."***(4) pags, 120-1

This excellent advice was given, to remind us,
to keep an open mind (Modified super-ego)*,
for the continuous new developments
in our respective fields of work.

The University DEGREE -vs- PRE-POTENCY and
The INTER-PERSONAL COMMUNICATIONS

The following Medical Doctors and Psychologists:
(Kurt Hauss, Peter Dentler, Norbert Neidenbach,
Herman Stegemann and Heinrich Volkel) explain the:
human mind and the good techniques for
Inter-personal communications,
and they mention that "It is true:
that everybody can communicate with another Person,
but it is very rare:
to know how to conduct or to have a:
successful professional dialogue."*(13)

"It is a very generalized mistake, to believe, that
a University degree
will automatically teach a person, how to obtain a:
successful positive reaction from another human being.

No University degree gives such guarantee or capacity,
because, it requires specific:
Knowledge, practice, and experience
{Modify the super-ego*and the Emotional Intelligence}
to fully develop
successful and satisfactory techniques for:
Inter-personal communications."*(13)

*(13) Fundamentals of Medical Psychology, by
Kurt Hauss, Publisher: Herder; Pag. 544.

Cont Section I: How To Manage "Yourself" (Soft Skills):

Therefore, to be able:
To modify our personality,
We must be:

Free of :
- Erroneous Super-Ego*:
  - Beliefs or Values or Habits or Knowledge
- Addiction to:
  - Substances or Things or Persons

To Be Mentally RECEPTIVE,
To be Able To: Listen & Understand,
One has To Be Free From Erroneous:

Pre-Judgements
&
Pre-Programming (Super-Ego)*

According to the world famous psychiatrist
Dr. Emilio Mira y Lopez, MD:
*The super-ego should be called Anti-ego, because
There is a sector, of psychological forces hostile to the Ego.*(5)pag202

The super-ego is the psychological **or Social environment,**
where we: Grew, lived, studied and **Imitated** the
Behaviors and Values: ethical, moral, religious and cultural.

Cont Section I: How To Manage "Yourself" (Soft Skills):

The brilliant ex-director of ITT, Mr. Harold Geneen,
Mentions in his book: "Managing"*(2)pag.185, that:
If we eradicate the "Egotism"** in "some" companies
{Due to un-modified super-ego* and Emotional Intelligence}
We can :

$$\text{Improve} \atop {40\% \atop \text{in:}} \left\{ {\text{Performance,} \atop {\text{Productivity} \& \atop \text{Earnings}}} \right\} {\text{Of any Contractor} \atop \text{or Company"}*(2)}$$

Thus, it is very important for professionals
To Improve how to manage themselves, to obtain:
   a) An excellent {Emotional + Social} Intelligence
   b) To cultivate the inner peace or tranquility
   c) To be able, to apply: The golden rule of psychology:
         "Be very sincere with yourself",
      To modify all the deficiencies of the super-ego,
   d)   To be able to achieve sustainable success at:
         The Professional, Emotional and Financial level.

Mr. Geneen also mentions a very wise advice:
"More careers are ruined by success than by failure"
{Due to an un-modified super-ego, because sometimes
The success goes way over their heads, and
Frequently they start mistreating others or themselves}.

To avoid this, it is advisable,
To modify the super-ego* and the Emotional Intelligence,
"To Be very sincere with yourself"
To be able, to follow the practical advice given by
The world renowned professionals, mentioned in
Sections I & II of this practical book;
To continue achieving sustainable success at:
The Professional, Emotional and Financial level.

According to the world famous, Dr. Emilio Mira y Lopez,MD;
*"The super-ego should be called Anti-ego, because sometimes,
There is a sector, of psychological forces hostile to the Ego."*(5)pg202

The super-ego is the psychological or **Social environment,**
      Where we: Grew, lived, studied and **Imitated** the
Behaviors and values: ethical, moral, religious and cultural.

**Egotism = Arrogant, or abusive or disrespectful Behaviors.

*(2) Managing, By: Harold Geneen, Pag185, Pub Avon Books.

Cont Section I: How To Manage "Yourself" (Soft Skills):
Matching points (*=*) of the **Management styles** of:
*Segundo Cristancho J. & Company {during 67 years}, and
*"Mr. Harold Geneen', Ex-CEO of ITT
{Helped a **$766** million telephone company grow into a
$22 Billion multinational conglomerate, according to
His book: Managing, Pub. by Avon Books}:*(2)
Several executives, that worked with Mr. Geneen,
graduated to successful careers as Fortune 500 CEOs:
*=*1) **Plan, direct and control:** "Know every facet of the businesses" and
= 2) Treat people with respect & admiration and listen with empathy,
have an open door, but avoid incompetence and neglect.
*=*3) Urgently, If present, modify the egotism,**
To be able, "To improve or regain 40% of the:
Quality, safety, performance, productivity,
Innovation, competitiveness and earnings of any
Corporation, contractor or profession
{Mr. H. Geneen, Ex-CEO of ITT}."*(2)".
{Modify the super-ego* and the Emotional Intelligence,
applying sections I & II of our practical book}.
*=*4) "Update the plans, associates must update them too.
Use what worked, but changed when necessary."
*=*5) "Be a: {Hands on, present, active and a responsible} manager."
*=*6) "Know:{your business, what works, your people, who works}
*=*7) "Manage with respect & admiration (w/o egotism)."**
"Performance and results must be achieved,
If you don't achieve those results, you're not managing"
{To Improve your management, apply sections I & II of our book}.
*=*8) "Put in the time and dedication to work things out."
Never delegate the responsibilty of controlling a
Task or a goal, to be done, with excellent:
Quality, safety, on time and under budget.
*=*9) "When a client mentions a suggestion or complaint,
mention it to management ASAP, and fix it asap,
according to the plans, contract and specifications,
to comply with the client wishes and desires; thus
always having a satisfied and loyal client."

*Also, applied at: **Segundo Cristancho J. & Co.** during 67 years.
*(2)Managing: Harold Geneen, ex- CEO, ITT, Pub. by Avon Books..
**Egotism = Arrogant, or abusive or disrespectful Behavior

GREAT MINDS ARGUE ABOUT **IDEAS,**
AVERAGE MINDS ARGUE ABOUT **THINGS,** and
LITTLE MINDS ARGUE ABOUT **PEOPLE** !!!

Las mentes grandes discuten ideas;
las medianas cosas; y las pequeñas,
personas.

Proverbio Chino

Cont Section I: How To Manage "Yourself" (Soft Skills):

Remember when you were born,
Everybody was laughing, and you were crying;
Then, live your life in such a way,
That when you die,
Everybody will be crying,
And you will be laughing !!!

Recuerda que el día en que naciste odos reían y tu llorabas;
Vive de tal manera que cuando mueras,
todos lloren y tu rías.

Proverbio Chino

Cont Section I: How To Manage "Yourself" (Soft Skills):

In the previous pages,
Worldwide renowned professionals
Described many ways showing:

How { To Modify or / Trans-form or / Re-Program } The Personality: / The Super-Ego* & / The Emotional Intelligence

And for this, we will need an:

Un-Hurried, / Calm / "Mind" / Capable of: { Listening, / Understanding and / Having a: / Flexible Super-Ego* }

Warning: This book is not intended to take the place of medical advice from a trained medical professional; thus, always seek the best medical assistance in ALL cases regarding your total health. Neither the authors, nor the publisher of this book, nor the publishers of the books mentioned in this book take any responsibility if this warning is not followed completely.

Cont Section I: How To Manage "Yourself" (Soft Skills):

In the previous pages, we defined and explained
some very important topics, to make it easier,
To understand yourself better,
To modify the personality
{Modify the super-ego* and the Emotional Intelligence}.*

It is very important to remember a statement made by
The famous psychologist Dr. Anthony (Tony) De Mello:
"Just concentrate on looking, and
Someday you will be able to see clearly."

In other words, from now on,
ANALYZE, COMPARE & APPLY
Everything that you read and:
**BE VERY SINCERE
WITH YOURSELF.**

This is the golden rule of psychology:
to achieve sustainable success at
The Professional, Emotional and Financial level.

The Family's psychologists have proven that:
Our lives are molded by transactions with our
Social or Psychological environment = Super-Ego,
Where we grew and lived; consequently,
We must observe and analyze our
Social or Psychological environment = Super-Ego,
To be able to modify our personality.

Our Personality
(Super-Ego)
Is formed during
Our Childhood and
Adolescent
}
And sometimes, the
Consequences of our
Mature life is the
Outcome of our experiences
During our: Childhood and
Adolescent

The "FAMILY"
Is one of the
Components of
the Super-Ego*,
and it :
{ **MOLDS**

&

MODELS }
ITS MEMBERS

Warning: This book is not intended to take the place of medical advice from a trained medical professional; thus, always seek the best medical assistance in ALL cases regarding your total health. Neither the authors, nor the publisher of this book, nor the publishers of the books mentioned in this book take any responsibility if this warning is not followed completely.

OUR CHILDREN'S EDUCATION

The children's Super-Ego*=
*The child's Super-Ego is the **Social** or psychological environment
Where he: Grows, lives, studies and **Imitates** the surrounding
Behaviors and **values:** ethical, moral, religious and cultural.

The world famous Dr. Emilio Mira y López in his book,
in his book, Guide to Mental Health recommends:
Healthy mind in a Healthy {Body and Society}; thus,
It is advisable, to give our children a lot of:

{ Love, Attention, Understanding, and
**Excellent VALUES and BEHAVIORS
To Imitate.**

But we must not spoil our children, so
It is advisable to establish:
{ **Material &**
Emotional } Limits

To avoid or prevent the child or adolescent to become a
Spoiled immature adult,
Incapable and ill equipped, to face and solve the
real life's problems in the adult and mature world, and
To avoid the child or adolescent future:
Financial and Emotional setbacks.

According to the world famous psychiatrist,
Dr. Emilio Mira y Lopez, MD:
"The super-ego should be called Anti-ego, because,
There is a sector, of psychological forces hostile to the Ego."*(5)pag202.

The super-ego is the psychological or **Social environment**
Where we: Grew, lived, studied and **Imitated** the environmental
Behaviors and Values: Ethical, moral, religious and cultural.

Cont Section I: How To Manage "Yourself" (Soft Skills):

In the previous pages, we have described:

HOW TO MODIFY THE
PERSONALITY
"SUPER-EGO"

{Modify the Super-Ego* and the Emotional Intelligence}.
Therefore, we recommend to review these previous pages,
so that it will be easier,
To modify the personality or the type "A" personality.

"The most significant trait of type "A" personality,
is the habitual "sense of urgency" or the
"Hurry Sickness",
this is the continuous fight of the type "A" person
against **time,** which causes, sometimes premature:
Cardio-coronary illnesses."*(4)

According to the world famous psychiatrist,
Dr. Emilio Mira y Lopez, MD:
"The super-ego should be called Anti-ego, because,
there is a sector, of psychological forces hostile to the Ego."*(5)pg202

The super-ego also defines the psychological or **Social environment,**
where we: Grew, lived, studied and **Imitated** the
Behaviors and Values: ethical, moral, religious and cultural.

*(4) Treating Type "A" Behavior & Your Heart, by
Meyer Friedman, MD & Diane Ulmer, RN, MS.,
Published by Fawcett Crest, NY.

CAN WE SCULPT OUR OWN BRAIN ??
YES,
According to Nobel Prize winner in Medicine,
Dr. Santiago Ramón y Cajal:
By TALKING TO OURSELVES,
We MOLD our EMOTIONS,
That MOLD our:
PERCEPTIONS and our LIVES !!!

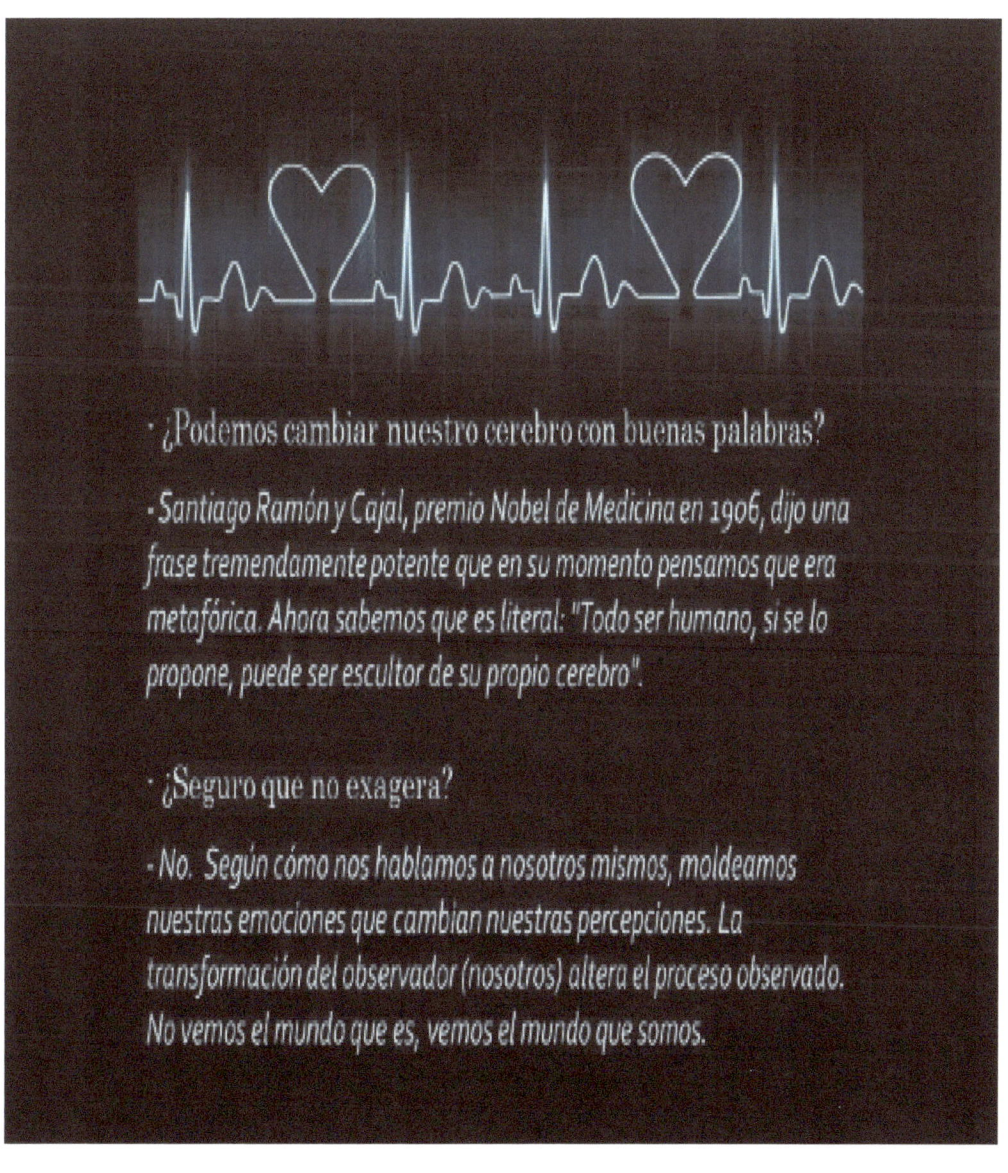

· ¿Podemos cambiar nuestro cerebro con buenas palabras?

· Santiago Ramón y Cajal, premio Nobel de Medicina en 1906, dijo una frase tremendamente potente que en su momento pensamos que era metafórica. Ahora sabemos que es literal: "Todo ser humano, si se lo propone, puede ser escultor de su propio cerebro".

· ¿Seguro que no exagera?

· No. Según cómo nos hablamos a nosotros mismos, moldeamos nuestras emociones que cambian nuestras percepciones. La transformación del observador (nosotros) altera el proceso observado. No vemos el mundo que es, vemos el mundo que somos.

Cont Section I: How To Manage "Yourself" (Soft Skills):

The complete understanding of yourself
Is obtained only, when there is a:
"Sincere" intention of understanding ourselves,
When the mind is:

Un-Hurried,
     Quiet,   CONDEMNATION,
     Serene,
     Tranquil,
     Positively Alert: JUSTIFICATION or
Free { Perturbation &
From: { Distraction
     and Without: IDENTIFICATION

This is one of the possible ways,
To understand ourselves,
To Modify our:
Personality
(Super-ego* = Psychological or **Social Environment**)
and to find out the real truth of:
Who we really are,
To be able to achieve sustainable success at
The Professional, Emotional and Financial level.

Cont Section I: How To Manage "Yourself" (Soft Skills):

The life of some people is full of some activities that
Confused the mind and
Diminish the vital energy.

To feed the sensations does **not** mean:
The agitated search of sensations,
It does **not** mean to travel to expensive & exotic places;
the real meaning of feeding the sensations implies:
To live life fully, covering all our surroundings.

The Human   Understands   Sentiments &
Being when   The Real: Thoughts
is content   &
and Happy: Attributes the real meaning to
                    what really counts.

Warning: This book is not intended to take the place of medical advice from a trained medical professional; thus, always seek the best medical assistance in ALL cases regarding your total health. Neither the authors, nor the publisher of this book, nor the publishers of the books mentioned in this book take any responsibility if this warning is not followed completely.

Cont Section I: How To Manage "Yourself" (Soft Skills):

According to psychologist Dr. G. E. Salesman,
he defines the following stages for human life cycle,
To reach psychological maturity :

CHILDHOOD   =   Up until 12 years
ADOLESCENCE   =   Up until 25
ADULTHOOD   =   From 25 to 50 years of age
MATURITY   =   From 50 to 70

When we reach   { Understand }   Reality,
  Psychological   The way it
Maturity, then,   &   Really works
  We begin to: { Accept

Maturity helps   { Pre-Potence,
Us to acknowledge   Arrogance,
The limits of   Omnipotence,   Fragile
Our true capacities, Greatness,   Super-Ego
To demolish our   Typical of the
Myths of :   { Persons with a:

Warning: This book is not intended to take the place of medical advice
from a trained medical professional; thus, always seek the best medical
assistance in ALL cases regarding your total health. Neither the authors,
nor the publisher of this book, nor the publishers of the books mentioned
in this book take any responsibility if this warning is not followed completely.

To obtain an adequate "maturity" development,
this is not achieved merely by deciding consciously,
that we are going to abandon the illusory dreams of
Prepotence, instead of that, we must:

Accept & ⎫ Our relationships with the rest
Enrich ⎭ of our fellow human beings.

The best type of ⎰ Flexible ⎱ That listen
Friendships   ⎱   &   ⎰ Without judging.
Are those that are: ⎱ Expressive ⎰

Sometimes, some problems in life will dis-appear
The next day, or they will be forgotten the next day,
or the intensity of the problem will be reduced in a
few days, so try not to exaggerate the intensity of
the problems, always remember:
On many occasions,
Time will help to solve or diminish
The intensity of many problems !!!.

Warning: This book is not intended to take the place of medical advice from a trained medical professional; thus, always seek the best medical assistance in ALL cases regarding your total health. Neither the authors, nor the publisher of this book, nor the publishers of the books mentioned in this book take any responsibility if this warning is not followed completely.

From the previous pages we can conclude that,
It is very important
Not to confuse your real self
with your false or imaginary self or ego, or
With the self or ego of somebody else.

Remember the advice of the world famous,
Dr. Emilio Mira y López, MD, who mentions that:
It is advisable to start improving oneself, by
Comparing ourselves against
"Our own" strengths and weaknesses,
Instead of comparing ourselves against
Others' strengths and weaknesses.

To Stop comparing yourself against others,
It is advisable, to follow the advice of
the world famous professionals,
Mentioned in sections I & II of this practical book;
So that it will be easier,
To modify the personality,
Modifying the super-ego* and the Emotional Intelligence,
To obtain Emotional and financial success.

Warning: This book is not intended to take the place of medical advice
From a trained medical professional; thus, always seek the best medical
Assistance in ALL cases regarding your total health. Neither the authors,
Nor the publisher of this book, nor the publishers of the books mentioned
in this book take any responsibility if this warning is not followed completely.

The "FALSE" EGO:

It is one thing what a person really is, and
another different thing is what that person
Think he is = "False" Ego of that person.

Reality is what a person really is, and an
Imaginary reality is:
What that person falsely imagines the real self to be.

When the reality is what a person really is,
then, we are very wise, because,
we are not deceiving ourselves,
Neither are we deceiving others.

When the ego does not accept reality,
then, that person suffers a complex of:
Pre-potence; thus, it is advisable,
To apply the golden rule of psychology:
"Be very sincere with "yourself"
To modify the super-ego* and the Emotional Intelligence, and
ANALYZE, COMPARE & APPLY
The advice of the world renown professionals
Mentioned in Sections I & II of this practical book.

According to the world famous,
Dr. Emilio Mira y Lopez, MD:
"The super-ego* should be called Anti-ego, because sometimes,
There is a sector, of psychological forces hostile to the Ego."*(5)pag202

The super-ego also defines the psychological or **Social environment,**
Where we: Grew, lived, studied and **Imitated** the
Behaviors and Values: ethical, moral, religious and cultural.

**Egotism = Arrogant, or abusive or disrespectful Behaviors.

Cont.   The "FALSE" EGO:

When a person is confused and identifies
with his/hers False Ego, then, that person
confuses what he/she really is,
With what he/she falsely
Imagines him/her self to be or to become someday;
Then, that person has a:
Floating or false personality, with a:

Desires,
Fears ,
False Ego,   Anxiety,
Full of   Obsessions or
False   Things or
Illusions of: Addictions   Persons or
               To :   Substances

"Sometimes" 30% of the
Financial and/or Emotional failures,
Could have been prevented or avoided
If they had not confuse their "real" self,
with their "imaginary or false" self; thus,
Avoiding or preventing
Themselves or others into:
Financial and /or emotional failures !!.

Warning: This book is not intended to take the place of medical advice
From a trained medical professional; thus, always seek the best medical
Assistance in ALL cases regarding your total health. Neither the authors,
Nor the publisher of this book, nor the publishers of the books mentioned
in this book take any responsibility if this warning is not followed completely.

We have to
Understand
Ourselves
To Know :

} WHO we really **ARE,**

WHAT we really **WANT** and

WHAT we are really **AFRAID** of

The Modification or
Re-Programming
of our Mind,
Is obtained by:
Understanding &
Comprehension
of our:

SENSES: {
Sight,
Smell,
Touch &
Hearing

&

PERCEPTION

Discovering
Our own: {
Desires,
Thoughts and
Persistence

Warning: This book is not intended to take the place of medical advice from a trained medical professional; thus, always seek the best medical assistance in ALL cases regarding your total health. Neither the authors, nor the publisher of this book, nor the publishers of the books mentioned in this book take any responsibility if this warning is not followed completely.

The beginning
of the:

{
Modification or
Trans-formation or
Re-programming
of our:
}

Personality (character)
Modify the Super-Ego

Is obtained:
Understanding better
The Super-Ego
of :

{
Our selves
&
Others
}

Remember what the :
world famous **Dr. Erich Fromm, MD,**
mentions in his book,
"The Art of Listening" Pags. (68,69):
"Only a Fundamental transformation of our
Personality system,
can produce a significant change of our character;
in other words:
we must change not only in one aspect,
but in all our aspects of our personality system;
that is, the way we:
Think, act, feel, move .... and everything else,
because, **one isolated Emotional change,**
never produces a lasting effect."
{Modify the super-ego* and the Emotional Intelligence}

According to the world famous,
Dr. Emilio Mira y Lopez, MD:
"The super-ego* should be called Anti-ego, because sometimes,
there is a sector, of psychological forces hostile to the Ego."*(5)pag202

The super-ego is the **Social or** psychological **environment,**
where we: Grew, lived, studied and **Imitated** the
Behaviors and Values: Ethical, moral, religious and cultural.

To: { **MODIFY or RE-PROGRAM OURSELVES,**

The world famous, Dr. Anthony de Mello mentions that:
"We have to take a chance without formulas,
discover everything ourselves,
observe & understand "anew" everyday:
all the processes & movements of life."

In other words, just:

"Look,
Observe   & { Everything
Around you } You will
Understand   and soon   Be able to see."

Warning: This book is not intended to take the place of medical advice from a trained medical professional; thus, always seek the best medical assistance in ALL cases regarding your total health. Neither the authors, nor the publisher of this book, nor the publishers of the books mentioned in this book take any responsibility if this warning is not followed completely.

Internal Self-Evaluation

NOTE: Always remember the golden rule of psycho-analysis:
"Be very sincere with yourself";
To be able, to perform a sincere Internal Self-Evaluation:

1) "Determine how intelligent and creative at your job.
Examine your sense of humor and **flexibility**:
to modify your personality (super-ego)."*(4)

2) "Examine all the "activities" that now absorb your:
Intellectual, emotional, and spiritual interests.
How many of these activities have to do with
with your concern for: art, literature, music,
drama, philosophy history, science??."*(4)

3) "Seek and assess your free-floating **hostility**."*(4)

4) "Estimate, the ease to receive and give :
Loyalty, and affection. "*(4)

5) "Determine the courage and fear you possess."*(4)

6) "Examine your ethical and moral principles.
How honest have I been in my life **?**."*(4)

7) "Finally, ask yourself what should be:
The essence of my life ?."*(4)

*(4) Treating Type "A" Behavior & Your Heart,
Pgs.(95,218,219,242) by: Meyer Friedman, MD &
D. Ulmer, RN, MS., Published by Fawcett Crest, NY.

Cont Section I: How To Manage "Yourself" (Soft Skills):

The University DEGREE -vs- PRE-POTENCY and
The INTER-PERSONAL COMMUNICATIONS

The following Medical Doctors and Psychologists:
(Kurt Hauss, Peter Dentler, Norbert Neidenbach,
Herman Stegemann and Heinrich Volkel)
Explain the:
Human Mind and the good techniques for
Inter-personal communications, and
They mention that "It is true:
That we communicate with each other,
But it is very rare:
To know how to conduct or to have a :
Successful professional dialogue."*(13)

"It is a very generalized **mistake,** to believe, that
A University degree
will automatically teach a person,
how to obtain a successful positive reaction
From another human being."*(13)

"**No** University degree gives such guarantee
or capacity, because, it requires specific:
Knowledge, practice, and experience
To fully develop:
Successful and satisfactory techniques for:
inter-personal communications,"*(13)
{Modify the super-ego* and the Emotional Intelligence}.

*(13): Fundamentals of Medical Psychology, by
Kurt Hauss, Publisher: Herder; Pag. 544.

Cont Section I: How To Manage "Yourself" (Soft Skills):

Cont. The University DEGREE -vs- PRE-POTENCY
And the INTER-PERSONAL COMMUNICATIONS

It will be quite appropriate, to mention the words of
wisdom of the world famous cardiologists:
Drs. Friedman, Rosenman, MDs, Ulmer & Associates:
**"We modern physicians and nurses
still have much to learn, and
much to be modest about."***(4) pags, 120-1

This excellent advice was given, to remind us,
to keep an open mind (flexible super-ego)*,
For the continuous new developments
In our respective fields of work.

This advice was mentioned by:
Dr. Friedman and associates,
due to the new developments, that they found
during twenty (20) years of research regarding
the fatal consequences of
The Un-modified Type "A" personality.
{Modify the super-ego* and the Emotional Intelligence}

According to the world famous psychiatrist,
Dr. Emilio Mira y Lopez, MD:
*"The super-ego should be called Anti-ego, because
There is a sector, of psychological forces hostile to the Ego."*(5)pag.202

The Super-Ego is the **Social or** psychological **environment,**
Where we: Grew, lived, studied and **Imitated** the
Behaviors and Values: ethical, moral, religious and cultural.

*(4)   Ibid : Pags(120, 121)

Cont Section I: How To Manage "Yourself" (Soft Skills):

Cont. The University Degree -vs- PRE-POTENCY and
The Inter-Personal COMMUNICATIONS

It will be very helpful, at this time, to remember
the famous and wise commentaries of the:
Asian philosophers,
Who mention the following advice:
The **wise** man looks for knowledge;
The Un-wise man thinks:
That they already have that knowledge.

Also, Dr. M. Teutsch, MD,
Mentions the following advice:
"It does not matter the amount of:
University degrees that you have obtained,
You will **not** be completely educated,
Until you have:
Modified your personality,
That is, increasing your:
Self-esteem and
Knowing yourself quite well"*(14)
{Modify the super-ego* and the Emotional Intelligence}.

According to the world famous,
Dr. Emilio Mira y Lopez, MD:
"The super-ego should be called Anti-ego, because,
There is a sector of psychological forces hostile to the Ego."

The super-ego is the **Social or** psychological environment,
Where we: Grew, lived, studied and **Imitated** the
Behaviors and Values: Ethical, moral, religious and cultural.

*(14) J.M. & C. Teutsch: From Here To Greater Happiness.

91

Cont Section I: How To Manage "Yourself" (Soft Skills):

Influence of SOCIAL or ENVIRONMENTAL Factors
(The Super-Ego) on the PERSONALITY :

"According to the psychological researcher:
Dr. Eysenck: The human behavior of a person,
depends on two (2) components:

$\begin{cases} \textbf{1) } \text{Constitutional and} \\ \textbf{2) } \text{Environmental,} \end{cases}$

and the following formula is given, to express the
components of the human behavior:

$$P.B = (G.F.) \times (E.S.) \qquad *(15)$$

Where :
P.B. = Personality Behavior,
G.F. = Genetic Factor of the Personality,
with respect to the response of the
CNS (Central nervous system) to the
environmental stimulus, such as:
excitement or inhibition from our
Sensorial capabilities.

E.S. = Environmental stimulus, which produces
Pleasure & Dis-Pleasure,

According to the world famous psychiatrist,
Dr. Emilio Mira y Lopez, MD:
"The super-ego should be called Anti-ego, because
there is a sector, of psychological forces hostile to the Ego."*(5)pag202

*The super-ego*is the psychological or **Social environment,**
Where we: Grew, lived, studied and **Imitated** the
Behaviors and Values: Ethical, moral, religious and cultural.
*(15) Ibid : Pag 456

Cont Section I: How To Manage "Yourself" (Soft Skills):

Influence of SOCIAL or ENVIRONMENTAL Factors
(The Super-ego) on the PERSONALITY:

P.B = (G.F.) x (E.S.) {See previous page}

This formula developed by Dr. Eysenck, shows that:
Personality Behavior (PB):
Is directly proportional to both the:
Genetic factor (GF) and also to
The Environmental or Social stimulus (ES) or super-ego;
Therefore, from now on,
It is advisable, to give high priority to these factors.

This Environmental or Social stimulus (ES) or
Super-ego describes the:
Teachings, examples, values, habits and customs
That we learned or tried to imitate from our:
Environment, Relatives, teachers or friends.

Perhaps, from this equation,
We have learned the popular saying:
Tell me who your friends are;
and then, I can tell you:
Who you are or will be !!!.

The good news is that by following
Sections I & II of this practical book, it will help you,
To modify the super-ego and the Emotional Intelligence;
To be able, to correct all the deficiencies of the previous
Social or Environmental stimulus (ES) or super-ego.

TO MODIFY The PERSONALITY (Super-Ego):
List of "possible" deficiencies of the Super-Ego,
That will be convenient to modify,
"To improve or regain 40% of the:
Quality, safety, performance, leadership,
Innovation, competitiveness and earnings of
The corporation, contractor, project or profession."*(2)

Drs. Friedman & Rosenman, MDs, Dr. Daniel Goleman, H.Geneen
and many CEOs, MDs, and scientists,
Recommend to modify the personality (character),
To apply the golden rule of psychology:
"Be very Sincere with Yourself,"
When analyzing which of the following symptoms needs to be modified,
To improve the Super-Ego and the Emotional Intelligence:

A) Easy to Diagnose Symptoms:
    1) Hyper-aggressiveness or hostility or
    2) Time urgency = Hurry sickness, or
    3) "Egotism = Arrogant, or abusive or disrespectful behavior, or
             Bad temper"*(2),*(4),*(5), or
    4) Problems with the Emotional Intelligence, to be able
           To listen and treat ourselves and others better;
           To avoid causing major: mistakes, accidents, or losses.

B) Hard to Diagnose Symptoms:
    1)   **a)** Self Esteem, and/or
       Un-Modified {   Insecurity (Inner Peace) and/or
                 b) Mental & } Development
                    Material }

           To avoid the:
    2) (Unconscious) Drive toward **"self-destruction"** of the:
        a) Career or business or Marriage and/or
        b) The personality or **the life.**

*(2) Managing, by H. Geneen (Ex-CEO ITT), Pag (184-5) Pub. by Avon Books.
*(4) Drs.Friedman & Rosenman, MDs and D. Ulmer: Treating type A Behavior & Your Heart,
    Pags (84,87,93,229) Published by F. Crest, N.Y., N.Y.
*(5) Dr. Emilio Mira y Lopez, MD, The four Giants of the Soul:
    Fear, Anger, Love and Duty pags 124,5, Edit. Lidiun

Definition of "Wealth":

There are two (2) components of Wealth:
1) **Inner** = Happiness = Satisfaction with life,
   Modified Super-Ego*and Emotional Intelligence
   Excellent Personality (Character)
         Inner peace, harmony = **60%**

2) **Outer** = Assets, Positions,
   Titles, Diplomas & money ($$$) = <u>**40%**</u>
                                    100%

} Wealth

Summary: Improving our **Inner** wealth,
Increases or preserves our **Outer** wealth.
Wealth is composed of two (2) elements:
1) The **Inner Wealth** = **60%** of our total wealth,
   It gives us Happiness = Satisfaction with life, inner peace,
   To be able, to apply the golden rule of psychology:
   "Be very sincere with yourself,"
To modify the Super-Ego,*and the Emotional Intelligence,
To obtain an excellent personality (Character), to be able:
   a) To avoid mistakes, accidents, or losses, and
   b) Improve our **Inner** wealth, thus,
      Increasing or preserving our **Outer** wealth, and also our:
         Mental, physical and financial Health.

2) The **Outer Wealth** = **40%** of our total wealth,
   Which is always exposed to the mortal trap, mentioned
   By the ex-President of ITT, H. Geneen: sometimes,
"More fortunes (or careers) **are destroyed by success than by failure,**
   Frequently, generating Egotism"**(2)
   = Arrogant, abusive or disrespectful behavior; sometimes,
      Causing harm to others or themselves.

Conclusion: If we utilize our improved **Inner** wealth,
To modify the Super-Ego* and the Emotional Intelligence; then,
We will be able, to increase or preserve our **outer** wealth, and also
Succeed at the emotional, professional and financial level.

According to the world famous, Dr. Emilio Mira y Lopez, MD:
*The super-ego = Anti-ego, because there is a sector, of psychological forces
Hostile to the ego. The super-ego is the **Social** or psychological environment,
where we: Grew, lived, studied and **Imitated** the Behaviors and
Values: ethical, moral, religious and cultural.

**(2) Managing, by Harold Geneen (Ex-CEO ITT), Pag (184-5) Pub. by Avon Books.

SECTION II
{Soft Skills}
{EMOTIONAL over INTELLECTUAL} INTELLIGENCE:

Section II:{EMOTIONAL over Intellectual} Intelligence
  "Emotional Intelligence (EI) is a combination of:
      1) Self Management and
      2) Learning Social skills in an excellent environment,
          To transform and optimize,
      The performance of the individual or the group."*(10)
The world famous Dr. Emilio Mira y Lopez, MD, mentions:
  "Healthy mind in a healthy {body and society}."*(17)

      On Nov. 1/202, Business Week mentioned that,
          "Some Professionals are:
  Hired for theirs excellent professional knowledge and
  Fired for theirs Un-modified **Emotional Intelligence**."
The following definition applies to all the professions pertaining to:
          Engineering & Construction:
  Contractors, Engineers, Architects, Inspectors and Unions.

      Definition of a **PROFESSIONAL**:
1) A true professional possesses two types of knowledge:

Types of   { 1) The Professional or practical wisdom, and
Knowledge: { 2) The **one self's knowledge** =
                      Modified: { Emotional Intelligence
                                      and Super-Ego;*
      To be able, To avoid mistakes, accidents or losses.

  2)   Always applies the golden rule of psychology:
              "Be very sincere with yourself,"
  To modify the Emotional Intelligence and the Super-Ego* pertaining to:
              "The Fear, Love, Duty and
  Anger: Hatred, enviousness, prepotency, bad temper, and egotism,**
  Mentioned by the world famous Dr. Emilio Mira y Lopez, MD
  {In his book, "The four Giants of the Soul", Edit. Lidiun}."*(5)

  3) A Professional continuously improves the **Inner** wealth;
      To be able, to improve or preserve the **Outer** wealth.

According to the world famous, Dr. Emilio Mira y Lopez, MD:
  *The super-ego = Anti-ego, because there is a sector, of
      Psychological forces hostile to the Ego.*(5)pag202.
  The super-ego is the psychological or **Social environment,**
Where we: Grew, lived, studied and **Imitated** the behaviors and
      Values: Ethical, moral, religious and cultural.
**Egotism = Arrogant, or abusive or disrespectful behaviors.
    *(2) Managing, by H. Geneen, Ex-CEO ITT, Pag (184-5) Pub. by Avon Books.
*(10) Special collections, The eading Teams with Emotional Intelligence, by
        Drs: Daniel Goleman, R. Boyatsis, A. Mckee, J. R. Katzenbach.
*(17) Dr. Emilio Mira y López, MD: Guia de la salud mental, pag 18, Edit. Oberon.

Drs. Friedman & Rosenman, MDs, Dr. Daniel Goleman,
H. Geneen and many MDs, psychologists, psychiatrists, and scientists,
Recommend, that to be able,
To modify the personality (character), and
The EMOTIONAL Intelligence,
It is advisable, to always apply the golden rule of psychology:
"Be Very Sincere with "Ourselves";
When analyzing which of the symptoms needs to be modified
To help modify the **Emotional Intelligence** and the Super-Ego.*
To be able, To improve or regain 40% of the:
Performance, productivity, leadership and earnings of any:
Company or profession.

Dr. Daniel Goleman,
A world famous Harvard psychologist, who made
The concept of Emotional Intelligence (EI) popular,
Published recently his new digital book:
The Brain and Emotional Intelligence (EI), New Insights;
Published by More than Sound.*(8)

Dr. Daniel Goleman described
EMOTIONAL Intelligence
as having four (4) elements:*(8)

SELF {
AWARENESS
&
MANAGEMENT

SOCIAL {
AWARENESS
&
SKILLS *(8)

*(8)Dr. Daniel Goleman: The Brain and Emotional Intelligence
New Insights, Published by More than Sound, 2011.
I recommend this important digital book.

*(8) Dr. Daniel Goleman, new book:
   The Brain and Emotional Intelligence, New Insights,
      Published by More than Sound.

Explanation of Emotional Intelligence and Its new insights:*(8)

	SELF	SOCIAL
RECOGNITION	Self Awareness :	Empathy:
	Emotional Awareness	Organizational Awareness
	Accurate Self Assessment	Understanding the Environment
REGULATION	Self Management	Social Skills
	Self Control :	Influence :
	Trustworthiness	Inspirational Leadership
	Conscientiousness	Developing others
	Adaptability	Influence
	Drive & Motivation	Building bonds
	Initiative	Team work and Collaboration

"Emotional Intelligence (EI) is a combination of:
  1) Self Management and
  2) Learning Social skills in an excellent environment,
        To transform and optimize,
     The performance of the individual or the group."*(10)
  "Healthy mind in a healthy {body and society}."*(17)pag18
{Achievable Modifying the Super-Ego*and the Emotional Intelligence}.

*Empathy:
     It is how well we Listen to the point of view of
     another person, putting ourselves in their shoes,
     to solve problems as a team member; thereby,
        Implementing an excellent Leadership role.

     Emotions: in Dr. Goleman's words,
  "Personal" competence, comes from being
        Aware of and regulating **one's own** emotions.
  "Social" competence is :
     Awareness and Regulation of **others'** emotions."*(8)

        Remember what the:
     world famous Dr. **Erich Fromm, MD,**
  Mentions in his book "The Art of Listening"Pags.(68,69):
  "Only a fundamental transformation of our personality system
     can produce a significant change of our character;
     It is advisable, **To change not only in one aspect,**
     But in all our aspects of our personality system
        that is, the way we:
     Think, act, feel, move .... and everything else,
        because, **one isolated emotional change,**
           Never produces a lasting effect."
  {Modify the Emotional Intelligence and the super-ego*}.

*(8) Dr. D. Goleman, The Brain and Emotional Intelligence:
        New Insights, Published by More than Sound, 2011.
*(10) Special collections, The eading Teams with Emotional Intelligence, by
        Drs: Daniel Goleman, R. Boyatsis, A. Mckee, J. R. Katzenbach.
*(17) Dr. Emilio Mira y López, MD: Guia de la salud mental, pag 18, Edit. Oberon.

The Brain and Emotional Intelligence, New Insights,
by Harvard's psychologist, **Dr. Daniel Goleman,**
I recommend this excellent digital book.*

*According to Dr. Daniel Goleman:
"One of the bad consequences of the un-modified (EI)
Emotional Intelligence is the dark side of it; called
The Dark Triad:
The Narcissist, Machiavellians and Sociopaths,
because they **lack** emotional **empathy;** thus,
these people are sociopaths, and
they do **not** care about the human consequences of
their lies or manipulations, and
they do **not** have regrets about inflicting cruelty."*(8)

"The sociopath's brain has deficits,
in several areas key to Emotional Intelligence:
the anterior cingulate, the orbitofrontal cortex,
the amygdala and insula, and in the connectivity of
these regions to other parts of the brain."*(8)

"Some of the consequences of these sociopaths are
recognizable in organizational life:
   1) The bullying
   2) The "Kiss-Up-Kick-Down" Boss
         who can be very charming to superiors,
            but abusive and tyrant to subordinates.
            Another is the embezzler, a crook.
   3) The freeloader, the person who always holds a
      coffee cup and never does a lick of work."*(8)

*(8) Dr. Daniel Goleman :
   The Brain and Emotional Intelligence, New Insights,
      Published by More than Sound, 2011.

**The book: The Brain and Emotional Intelligence,
New Insights,**
by Harvard's psychologist, **Dr. Daniel Goleman;**
I recommend this important digital book.*
*Dr. Goleman describes some of the functions of
some parts of the brain:

"The **anterior Cingulate:**
It is the area that manages impulse control,
the ability to handle our emotions."*(8)

"If the right **Somatosensory cortex** gets injured, it
creates a deficiency in self-awareness and **empathy.**
empathy also depends on the Insula, because,
the **Insula** tells us: How we are feeling ourselves,
to sense and understand
what someone else is feeling."*(8)

"The **Pre-Frontal Cortex:**
Located behind the forehead, and it is the
Last part of the brain to become fully grown,
this is the brain's **executive center;**
here reside the abilities:
to solve personal and interpersonal problems,
to manage our impulses, to express our feelings
effectively, and to relate well with others."*(8)

"**The right Amygdala** is responsible for our
Emotional self-awareness,
the ability to be aware of and understand
our own feelings."*(8)

*(8) Dr. D. Goleman: The Brain and Emotional Intelligence:
New Insights, Published by More than Sound, 2011.

The book: The Brain and Emotional Intelligence, New Insights,
by Harvard's psychologist, Dr. Daniel Goleman,
I recommend this important digital book.*

According to Dr. Goleman, to modify **"old" habits,**
"It takes (3-6) months of practice, before the
"New" habit comes more naturally than the **old."***(8)

Description of the Social Emotional Learning (SEL) programs:

Dr. Goleman mentions that, at some schools in the USA,
A new program called **SEL: Social Emotional Learning**
have been implemented,
To improve the Emotional Intelligence of students."

"On the wall of every classroom there is a picture of a
stoplight, with its red, yellow and green lights."*(8)

RED light means:
STOP = **Calm down, and think before you act;**
teaching students, that:
you can not control what you are going to feel,
but you can decide what to do next.

YELLOW light means:
Think of the range of things you might do,
and what their consequences would be; thus,
enabling the student to learn to pick the best alternative.

The **GREEN** light means =
Try it out, and see what happens,
this is drilled into the students, and it works;"*(8)
Thus, "achieving the main purpose of learning
How to cope with disruptive feelings, to be able
To get along better with other people, and
Improve their performance at:
School or work."*(8)

*(8) Dr. D.Goleman: The Brain and Emotional Intelligence:
New Insights, Published by More than Sound, 2011.

Special Collections
The Leading Teams with Emotional Intelligence;
Collection by Drs.: Daniel Goleman, Richard Boyatzis,
Annie McKee, Jon R. Katzenbach.*(10)
"Emotional intelligence (EI) is a combination of:
1) Self-management and
2) Social skills,
that can transform and optimize
individual or team performance."*(10)

"Three **(3)** conditions are essential to a group's effectiveness:
Trust, Group {Identity, and Efficacy}."*(10)

"The TEAM needs to create Emotionally Intelligent Norms:
1) The Attitudes and
2) Behavior **(Habits)** for building:
Trust, Group {Identity, and Efficacy}.
3) Be mindful of the emotions of its **members,**
Its "own" **group** emotions or moods, and
the emotions of "other" groups and
individuals "outside" its boundaries."*(10)

"The task processes that distinguish the most
Successful teams— are the need for:
1) Cooperation,
2) Participation,
3) Commitment to goals, and so forth."*(10)

*(10) Special Collections The Leading Teams with
Emotional Intelligence, Collection by: Drs: D. Goleman,
R.Boyatzis, A. McKee, J. R. Katzenbach.

*According to Dr. Daniel Goleman :

"**Group (team) Emotional Intelligence** is about, bringing
Emotions deliberately to the surface, and
Understanding How they affect the team's work.
It's also about, behaving in ways that build relationships,
both inside and outside the team, and that strengthen
The team's ability to face challenges."*(10)

"Emotional intelligence means exploring, embracing, a
New awareness and regulation of **emotions** directed
both **inward,** to one's self, and **outward,** to others.
A **team** with emotionally intelligent members,
Does **not** necessarily make for an
Emotionally Intelligent group (**team**)."*(10)

"A team, takes on its own character.
So to create **trust, group identity, and group efficacy,**
it requires a team atmosphere in which:
the norms build emotional capacity
{the ability to respond constructively,
in emotionally uncomfortable situations} and
Influence emotions in constructive ways."*(10)

"**Personal** competence" in Dr. Goleman's words, comes
from Being aware of and regulating one's own emotions.
"**Social** competence" is:
Awareness and Regulation of others' emotions."*(10)

"A group (Team) must be mindful of the emotions of its
members, its own group emotions or moods,
and the emotions of other groups and
Individuals **outside** its boundaries."*(10)

*(10) Special Collections The Leading Teams with EI, Collection,
By: Drs: D. Goleman, R. Boyatzis, A. McKee, J. R. Katzenbach.

Improving
The Emotional
and Social   { Moral,
Intelligence   Religious   } VALUES
Improves: and
                    Ethical

"If we improve our spiritual life, we will become rich:
     because, we never find a poor person
          with a rich spiritual life !!! *(11)

"The painful   } Are fountains of
 Experiences   } Knowledge"*(11)

"Because "sometimes" through suffering is
     the only way that the stubborn learns."*(11)

Because,
Suffering
makes people   { Wisdom   } "Sometimes" difficult
Think back   and   to obtain by any
   and   Maturity   other means"*(11)
Obtain better:

*(11) Pag 801, Comentarios B. LatinoAmericanos.
     A.T II; Editorial V.D.; España.

108

LIFE's Secrets:

The "Secrets" for Obtaining and Retaining : { FINANCIAL, PROFESSIONAL and EMOTIONAL } Sustainable WEALTH

" Lifes' Temptations: { To be somebody To have: } Power and Money*(11)

"Those temptations make people suffer tremendously, making life a continuous search toward God; that is why, human beings take their sufferings to church, where they find strength, to overcome and continue Life's struggles with greater strength and hope."*(11)

To be able to succeed in life, it is advisable:
"To Have God as a permanent partner."*(11)

"To Achieve Sustainable Wealth : { MATERIAL, SPIRITUAL, PROFESSIONAL, EMOTIONAL and FINANCIAL } Obtained by the one who: Listens, Understands and Applies the Advice of God"*(12)

*(11) Pag 732, Comentarios B. Latino Americanos.
A.T II; Editorial V.D.; España.
*(12) Pag 343, Comentarios B. LatinoAmericanos.
N.T; Editorial V.D.; España.

Washington D.C. Sept 22/2016,
Recommendations from pope Francis on his visit to the USA:
It is advisable to modify the
Excessive individualism, egotism, and indifference
Some persons have, regarding the rest of mankind.

The world famous Dr. Emilio Mira y Lopez, MD, mentions that:
"A Healthy mind in a healthy {body and society}."*(17)pag18
{Achievable Modifying the Super-Ego*and the Emotional Intelligence}.

To achieve "A healthy mind in a healthy {body and society},
Pope Francis also mentioned that,
We have to take care of our:
Children, grandparents and the family.

The children are the future of mankind, and
The grandparents are the memory bank,
To help raise the children, providing them,
with the best role models to imitate, with
Excellent Habits, behaviors and **Values:**
Ethical, moral, religious and cultural; thus,
Helping the family to remain:
United, alive and contributing to society's wellbeing.

"Emotional Intelligence (EI) is a combination of:
1) Self Management and
2) Learning Social skills in an excellent environment,
To transform and optimize,
The performance of the individual or the group."*(10)

*(10) Special collections, The eading Teams with Emotional Intelligence, by
Drs: Daniel Goleman, R. Boyatsis, A. Mckee, J. R. Katzenbach.
*(17) Dr. Emilio Mira y López, MD: Guia de la salud mental, pag 18, Edit. Oberon.

Remember how our mind works:
The thoughts of our Mind*
are converted into words,
and those words will become our
Behavior, Values and Habits;
which will define our
personality (Super-Ego) and our:
Success in life.

Our *Mind contains our:
Super-Ego = Moral censor or
Our psychological or **Social environment,** where we:
Grew, lived and developed in life and
this is where we:
Learned, copied and imitated our
Examples or models of:
Behavior, Habits and
Values: Ethical, Moral and Religious,
which will define our personality and our:
Success in life.

Consequently, remember, that to be able,
To modify your personality, It is advisable:
To ANALYZE, COMPARE & APPLY
Everything that you read and

BE VERY SINCERE WITH YOURSELF.

This is the golden rule of psychology,
that will help us
To modify the personality
{Modify the super-ego* and the Emotional Intelligence};
To achieve sustainable success at
The Professional, Emotional and Financial level.

Remember what the:
world famous psychiatrist **Dr. Erich Fromm; MD,**
mentions in his book
"The Art of Listening" Pags.(68,69):
"Only a fundamental transformation of our personality system
can produce a significant change of our Character;
in other words:
we must change not only in one aspect,
but in all our aspects of our personality system;
that is, the way we:
Think, act, feel, move .... and everything else,
because, **one isolated emotional change,**
Never produces a lasting effect."
{Modify the Super-Ego* and the Emotional Intelligence}

Consequently, it is imperative, that
we acquire more insight into our own:
Strengths and Weaknesses,
To be able to modify our personalities, and
Achieve success at the:
Professional, Emotional and Financial Level.
{Modify the Super-Ego* and the Emotional Intelligence}

The book: *"Snakes in Suits": When Psychopaths Go to Work, by Drs. R. D. Hare and P. Babiak, Phds.*(7), mentions that: Some of the red flags of the psychopaths might be:

1) Superficially charming,
2) Pre-potent, and/or narcissists,
3) Lie & manipulate people easily,
4) **Lack remorse and empathy,**
5) Are cold, in-considerate and
     Mean, and do not accept responsibility.
6) Only care about: Money & Position,
7) Are irritables, impatients & impulsives: (the 3 I)
8) Are Bullies/or abusive with their subordinates,
     But very nice to their superiors,
     Thus, creating:Low morale or
     Loss of: (productivity or satisfaction) or
          Market share & Earnings,
     and also high personnel turnover"*(7)

"Dr. S. Freud tried unsuccessfully to explain."*(8)
     Why the psychopath has not yet resolved,
some very important unconscious childhood's issues
{relating to the unresolved unconscious hostility toward one of the parents}.

According to Dr. S. Freud,
     The child un-consciously hates one of the parents;
Thus, lacks that important parental censorship activity, and
     Continues braking all the norms or duties taught,
     Lacking - according to Dr. S. Freud -
     the so called Super-Ego {or Moral Censor}"
{All this information, is mentioned by the world famous psychiatrist,
     Dr. Emilio Mira y Lopez, MD, in his book,
          "The four Giants of the Soul"
"The Fear, Anger, Love and Duty", pag. 201-2, publ. by Lidiun}.

"Sometimes the Psychopath has Brain's abnormalities in the sections of the Ventrolateral, orbitofrontal Cortex and Amygdala."*(8)

*(8)The Brain and Emotional Intelligence: New Insights, By Harvard's Psychologist, Dr. Daniel Goleman,Published by More than Sound, 2011. I recommend this Important digital book.

SECTION III.

EXECUTIVE SUMMARY:

Contractors, Engineers, Architects, Inspectors & Unions, "Can Improve or Recuperate 40% of the: Quality, safety, performance, leadership, Innovation, competitiveness and earnings of any Corporation, Contractor or profession."*(2)

*(2) Managing, by Harold Geneen (Ex-CEO ITT), Pag (184-5) Pub. by Avon Books.

According to world renowned management adviser
Dr. Peter Drucker: "Managers should indeed know more about
Human beings, above all, managers and
{Professionals = Contractors, Engineers, Architects, Inspectors and Unions}
Need to know much more about "Themselves" (soft skills),**
so that they will not impair performance."*(1)

In other words: **"Know Thyself,"***(1)
to be able, to improve:
How we manage **"Ourselves"**,
Before we can manage **"Others"** successfully, and
"Improve or regain 40% of the
Quality, safety, performance, leadership,
Innovation, competitiveness and earnings of any
Corporation, contractor or profession
{Mr. H. Geneen, Ex-CEO of ITT}"*(2)
{Advisable to modify the Super-Ego* and the Emotional Intelligence (E.I.)}.

Consequently, it is advisable,
"To Be very sincere with "yourself",
when analyzing the symptoms to be modified;
To improve the super-ego* and the Emotional Inteligence (E.I.),
To be able, to achieve sustainable success, at
The Professional, emotional and financial level.

Note: On Nov. 11/2002, Business Week mentioned that:
"Some Presidents (or Professionals) are:
Hired for their excellent Professional Knowledge and
Fired for theirs Personalities"= Un-Modified super-ego* and E.I.

Soft Skills = Intelligence {Emotional + Social}
Hard Skills = Knowledge {Professional, Scientific + Practical}

According to the world famous,
Dr. Emilio Mira y Lopez, MD:
The super-ego should be called anti-ego, because,
there is a sector, of psychological forces hostile to the Ego."*(5) pag.202
The super-ego = The psychological or **Social environment**,
where we: Grew, lived, studied and **Imitated** the
Behaviors and Values: Ethical, moral, religious and cultural.
**Egotism = Arrogant, or abusive or disrespectful Behavior.

*(1)Management:Tasks, Responsibilities and Practices by P. Drucker, Pag. 244, Pub. Harper & Row
**(2) Managing By: Harold Geneen, Pg 185, Pub. by Avon Books.

Matching points (*=*) of the **Management styles** of:
*Segundo Cristancho J. & Company {during 67 years}, and
*"Mr. Harold Geneen, Ex-CEO of ITT
{Helped a **$766** million telephone company grow into a
$22 Billion multinational conglomerate, according to
His book: Managing, Pub. by Avon Books}:*(2)
Several executives, that worked with Mr. Geneen,
graduated to successful careers as Fortune 500 CEOs:

*=*1) **Plan, direct and control:** "Know every facet of the businesses" and

= 2) Treat people with respect & admiration and listen with empathy,
have an open door, but avoid incompetence and neglect.

*=*3) Urgently, If present, modify the egotism,**
To be able, "To improve or regain 40% of the:
Quality, safety, performance, productivity,
Innovation, competitiveness and earnings of any
Corporation, contractor or profession
{Mr. H. Geneen, Ex-CEO of ITT}."*(2)".
{Modify the super-ego* and the Emotional Intelligence,
applying sections I & II of our practical book}.

*=*4) "Update the plans, associates must update them too.
Use what worked, but changed when necessary."

*=*5) "Be a: {Hands on, present, active and a responsible} manager."

*=*6) "Know:{your business, what works, your people, who works}

*=*7) "Manage with respect & admiration (w/o egotism)."**
"Performance and results must be achieved,
If you don't achieve those results, you're not managing"
{To Improve your management, apply sections I & II of our book}.

*=*8) "Put in the time and dedication to work things out."
Never delegate the responsibilty of controlling a
Task or a goal, to be done, with excellent:
Quality, safety, on time and under budget.

*=*9) "When a client reports a complaint or suggestion,
Communicate it to management ASAP, and fix it asap,
according to the plans, contract and specifications,
to comply with the client wishes and desires; thus
always having a satisfied and loyal client."

*Also, applied at: **Segundo Cristancho J. & Co.** during 67 years.
*(2)Managing: Harold Geneen, ex- CEO, ITT, Pub. by Avon Books.
**Egotism = Arrogant, or abusive or disrespectful Behavior

118

Cont Section III: Executive Summary
This book contains practical management wisdom,
from brilliant world renowned: Presidents (CEOs),
Professionals (Contractors, Engineers, Architects, Inspectors & Unions),
Universities, and MDs mentioned;
To help us Improve **How we Manage "Ourselves."**

It is advisable, to apply the golden rule of psychology
"Be very Sincere with yourself"
when analyzing, which symptoms need to be modified,
To be able, to modify the Super-Ego* and the Emotional Intelligence.

This book, will help us obtain, an
Excéllent personal management style,
"To improve or regain 40% of the:
Quality, safety, performance, leadership,
Innovation, competitiveness and earnings of any
Corporation, contractor, project or profession
{Mr. H. Geneen, Ex-CEO of ITT};"*(2) and also,
To achieve professional, emotional and financial sustainable success.

According to the world famous
Dr. Emilio Mira y Lopez, MD:
*The super-ego should be called Anti-ego, because
There is a sector, with psychological forces hostile to the Ego.*(5)pag202

*The super-ego also defines the psychological or
Social environment,
Where we: Grew, lived, studied and **Imitated** the
Behaviors and **Values**: Ethical, moral, religious and cultural.

Warning: This book is not intended to take the place of medical advice
from a trained medical professional; thus, always seek the best medical
assistance in ALL cases regarding your total health. Neither the authors,
nor the publisher of this book, nor the publishers of the books mentioned
in this book take any responsibility if this warning is not followed completely.

*(2) Managing, by H. Geneen (Ex-CEO ITT), Pag (184-5) Pub. by Avon Books..
*(5) Dr. Emilo Mira y Lopez, MD: The four Giants of the Soul, pag 202, Edit. Lidium

Cont Section III: Executive Summary
It is advisable, **To Analyze, Compare & Apply**
sections I & II of this practical book, to be able,
To Apply the wisdom of the world renowned: Presidents (CEOs),
Professionals (Contractors, Engineers, Architects, Inspectors & Unions),
Universities, and MDs mentioned.

To be able, **"To improve or regain 40% of the:**
Quality, safety, performance, productivity,
Innovation, competitiveness and earnings of any
Corporation, contractor or profession
{Mr. H. Geneen, Ex-CEO of ITT}."*(2)

It is advisable, to apply the golden rule of psychology:
"Be Very Sincere with "yourself";
when Analyzing, which symptoms need to be modified, to help
Modify the Super-Ego* and the Emotional Intelligence.

This way, the following definitions can be applied,
To improve the innovations or changes required,
for the competitiveness and survival of the:
Corporation, contractor, project or profession:
- **a)** How to modify the personality
- **b)** Egotism**
- **c)** An excellent Engineer or Inspector
- **d)** Wealth: $\begin{cases} \textbf{Inner = 60\%} \text{ \&} \\ \textbf{Outer = 40\%} \end{cases}$

According to the world famous,
Dr. Emilio Mira y Lopez, MD:
*"The super-ego should be called anti-ego, because sometimes,
There is a sector, of psychological forces hostile to the Ego."*(5)pag 202

*The super-ego also defines the psychological or
Social environment where we:
Grew, lived, studied and **Imitated** the
Behaviors and Values: Ethical, moral, religious and cultural.
*Super-ego = In psychoanalysis, is that part of the psyche which:
- **1)** Is critical of the self or Ego, and
- **2)** Enforces moral standards, at an unconscious level,
  {**If** we have modified all the deficiencies of the Super-Ego};
  then, it blocks the irrational unacceptable impulses of the **Id**.*

*The **Id** = Irrational: instincts, impulses, passions; controls the pain & pleasure.
**Egotism = Arrogant, or abusive or disrespectful Behavior.

THE "STRUCTURAL" COMPONENTS of
The "PERSONALITY":
1) The ID   2) The EGO   3) The SUPER-EGO

*The ID: {
INSTINCTS
IMPULSES
PASSIONS
} **Irrationals &**
Un-conscious

CONTROLS The (Pain & Pleasure)

The EGO: {
Perception of: Reality & Reason

Tries To Regulate & Control the ID
}

SUPER-EGO:*
{Moral Censor}
{
Habits,
Examples,
Beliefs,
Teachings
Behaviors &
Values,*
of Our:
} {
Parents,
Relatives,
Teachers,
Neighborhood,
Religious Advisers,
Friends
}

Social or Psychological Environment
Where we: Grew, lived, studied and **Imitated**
The **Behaviors & Values:**
Ethical, moral, religious and cultural.

According to the world famous psychiatrist,
Dr. Emilio Mira y Lopez, MD:
"The super-ego should be called Anti-ego, because
there is a sector of psychological forces hostile to the Ego."*(5)
*The super-ego also defines the **Social or** psychological environment
where we: Grew, lived, studied and **Imitated** the
Behaviors and **Values:** Ethical, moral, religious and cultural.

Super-ego = In psychoanalysis, is that part of the psyche which:
  1) Is critical of the self or Ego, and
  2) Enforces moral standards, at an unconscious level,
    {**If** we have modified all the deficiencies of the super-ego};
    then, it blocks the irrational unacceptable impulses of the **Id.***

*According to Dr. Daniel Goleman:
"One of the bad consequences of the un-modified
Emotional Intelligence is the dark side of it; called
The Dark Triad:
The Narcissist, Machiavellians and Sociopaths,
because they lack emotional empathy; thus,
these people are sociopaths, and
they do **not** care about the human consequences of
their lies or manipulations, and they
do **not** have regrets about inflicting cruelty."*(8)

"The sociopath's brain has deficits in
several areas key to Emotional Intelligence:
The anterior cingulate, the orbitofrontal cortex,
the amygdala and insula, and in the connectivity of
these regions to other parts of the brain."*(8)

"Some of the consequences of these sociopaths are
recognizable in organizational life:

1) The bullying
2) The "Kiss-Up-Kick-Down" boss
who can be very charming to superiors,
but abusive and tyrant to subordinates;
another is the embezzler, a crook."
3) The freeloader, the person who always holds a
coffee cup and never does a lick of work."*(8)

*(8) Dr. Daniel Goleman:
The Brain and Emotional Intelligence, New Insights,
Published by More than Sound, 2011.

To MODIFY The PERSONALITY (Super-Ego):
List of "possible" deficiencies of the Super-Ego,
that will be convenient to modify,
"To improve or regain 40% of the:
Quality, safety, performance, leadership,
Innovation, competitiveness and earnings of
The corporation, contractor, project or profession."*(2)

Drs. Friedman & Rosenman, MDs, Dr. Daniel Goleman, H.Geneen
and many CEOs, MDs, and scientists,
Recommend to modify the personality (character),
To apply the golden rule of psychology:
"Be very Sincere with Yourself,"
When analyzing, which of the following symptoms needs to be modified,
To improve the Super-Ego and the Emotional Intelligence:

A) Easy to Diagnose Symptoms:
1) Hyper-aggressiveness or hostility or
2) Time urgency = Hurry sickness, or
3) "Egotism = Arrogant, or abusive or disrespectful behavior, or
Bad temper"*(2),*(4),*(5),*(19) or
4) Problems with the Emotional Intelligence, to be able
To listen and treat ourselves and others better;
To avoid causing major: mistakes, accidents, or losses.

B) Hard to Diagnose Symptoms:
1)
Un-Modified { a)Self Esteem, and/or
Insecurity (Inner Peace) and/or
b) Mental & } Development
Material }

To avoid the:
2) (Unconscious) Drive toward "self-destruction" of the:
a) Career or business or Marriage and/or
b) The personality or the life.

*(2) Managing, by H. Geneen (Ex-CEO ITT), Pag (184-5) Pub. by Avon Books.
*(4) Drs.Friedman & Rosenman, MDs and D. Ulmer: Treating type A Behavior & Your Heart,
Pags (84,87,93,229) Published by F. Crest, N.Y., N.Y.
*(5) Dr. Emilio Mira y Lopez, MD, The four Giants of the Soul:
"Fear, Anger, Love and Duty" pags 124,5, Edit. Lidiun
*(19)The Real life MBA, by Jack & Suzy Welch, pags 145, Pub. Harper, NY.,NY.

Cont. Section III: Executive Summary

Some of the problems of: Quality, Safety and productivity
are caused by:

Lack of: { Knowledge or
Attention or
Care or rest,

But "Sometimes"
Problems of
Quality, safety and
Productivity are
Caused (5-10)%
Due to the
Following reasons:

{ Personal or marital problems or
Prepotency and/or narcissism or
Problems as a team member or
Addiction to: { Substances or
Alcohol or
Sex or Persons or
Lack Values* or pride in the work
Performed, producing perception
Differences of: { Quality, Safety
and Productivity.

*Values = Ethical, Moral, cultural and Religious.

Summary:
Besides working hard and knowing their job's scope,
It is advisable for all professionals:
Contractors, Engineers, Architects, Inspectors and Unions,
To follow the advice of world re-known
Management's adviser Peter Drucker:
"Managers and their associates should
know more about human beings,
above all, they all need:
To know more about "themselves,"
So that they will **not** impair performance."*(1)

In other words:
"Know and Modify Thyself"
{Advisable, To Modify the super-ego*and the Emotional Intelligence},
To modify the personality {character}
and achieve sustainable success at
the Professional, Emotional and Financial level.

*(1) Management Tasks, Responsibilities and Practices by:
Peter F. Drucker, Pag. 244, Pub. by Harper & Row.

Practical Applications of Good Emotional Intelligence:
Personal and Professional Management

To perform an excellent Personal and Professional Management analysis
It is always advisable to forecast
the three (3) probable case scenarios:
1) The Best
2) The "Medium" and
3) The Worst probable case scenario,
and the contingency plans for each scenario,
in case something goes wrong or unforeseen;
because an excellent professional with good
Soft skills: = {Emotional + Social} Intelligence
will be able to realize,
that any of those three (3) scenarios
can happen at any time
the best, medium or worst case scenarios;
and this is an indication of :
Excellent Emotional Intelligence at work !!!.

Also, when analyzing projects remember:
**If there is a chance that something will go wrong, it will,
Unless we take precautions to avoid or mitigate it;** also
If there are 10 good positive outcomes, and 2 negative outcomes,
frequently the 2 negative outcomes will happen first.
Thus, when performing the contingency plans, always
Utilize an excellent modified Super-ego*and Emotional Intelligence
for the three (3) case scenarios: the best, medium or worst.

Note: HOW **To Delegate** a Task or a Goal:
We can delegate a task or a goal to be achieved; **but
we can not delegate the responsibility of controlling that:**
The task or goal is being achieved with the **best:**
Quality, safety, on time and under budget; **thus,
the controlling responsibility
Always belongs to the contractor, the professionals & the team.**
{Utilize an excellent modified Super-ego*and Emotional Intelligence}

Practical Applications Relating to Emotional Intelligence:

Comments by Harvard's (HBS) weekly newspaper by
Jehan de Fonseka, editor-in-chief of The Harbus:
"First Test people for:
Emotional & Social Intelligence; Back to basics...
How well do they care about each other."

"The idea behind this is that:
Good **leadership** begins with **self-knowledge.**
The problem with "some" leaders today has
little to do with their ability to crunch numbers,
but rather a lack of Values"
{Modify the Super-Ego*and the Emotional Intelligence}.

"When you think about the:
biggest failures of corporate executives,
they are not necessarily technical failures,
But Ethical ones"
{Advisable To Modify the Super-Ego*and the Emotional Intelligence}.

"We need more insight into our own:
Strengths and Weaknesses,
with more training & practice in:
"Emotional Intelligence"
to get Back to basics....
to care more about each other,"
to be able, to achieve sustainable **success** at:
The Professional, Emotional and Financial level
{Advisable To Modify the Super-Ego*and the Emotional Intelligence}.

Type "A" Personality

"Most type "A" possess so much aggressive drive,
that it frequently evolves into a free-floating
Hostility".*(4)
"Perhaps, the prime index of the presence of
aggression or hostility in almost all type "A" persons,
is the tendency to always
To compete with or to challenge other people,
whether, the activity consists of a sporting contest,
a game of cards, or a simple discussion."*(4)

"There are some persons, whom we consider
type "A" because they are so **hostile,** that
They are almost continuously engaged in a:
Struggle against another person."*(4)

"Type "A" individuals do tend to seek each other
out socially, despite the fact, that often their free
floating Hostility and excessive competitiveness,
sometimes, converts their social meetings into
War meetings."*(4)"

*(4) Treating Type "A" Behavior & Your Heart, by
    Meyer Friedman, MD & Diane Ulmer, RN, MS.,
    Published by Fawcett Crest, NY., Pags.(95,227)

This definition applies to all the professions pertaining to:
Engineering & Construction:
Contractors, Engineers, Architects, Inspectors and Unions.

Definition of a **PROFESSIONAL**:

1) A true professional possesses two types of knowledge:

Types of knowledge
{
1) The Professional or practical wisdom, and
2) The **one self's knowledge** =

Modified: { Emotional Intelligence and Super-Ego;*

To be able, To avoid mistakes, accidents or losses.

2) Always applies the golden rule of psychology:
"Be very sincere with yourself,"
To modify the Emotional Intelligence and the Super-Ego* pertaining to:
"The Fear, Love, Duty and
Anger: Hatred, enviousness, prepotency, bad temper, and egotism,**
Mentioned by the world famous Dr. Emilio Mira y Lopez, MD
{In his book, "The four Giants of the Soul", Edit. Lidiun}."*(5)

3) Optimizes the relationships with others, to listen to them with
Empathy, respect and admiration; to implement the
Innovations and changes required,
To improve or recuperate 40% of the
Quality, safety, performance, leadership,
Innovation, competitiveness and earnings of the:
Country, corporation, contractor, project or profession.

4) A Professional continuously improves the **Inner** wealth;
To be able, to improve or preserve the **Outer** wealth.

On Nov. 1/202, Business Week mentioned that,
"Some Professionals are:
Hired for theirs excellent professional knowledge and
Fired for theirs Personalities" = Un-modified Super-Ego*and Emotional Intelligence.

According to the world famous,
Dr. Emilio Mira y Lopez, MD, the
*Super-Ego = Anti-Ego, because there is a sector, of
psychological forces hostile to the Ego.*(5)pag.202

The Super-Ego is also, the psychological or **Social environment**,
where we: Grew, lived, studied and **Imitated** the behaviors and
Values: ethical, moral, religious and cultural.
Egotism = Arrogant, or abusive or disrespectful behaviors.
*(2) Managing, by H. Geneen, Ex-CEO ITT, Pag (184-5) Pub. by Avon Books.

Consequently, in this book we used some of
The knowledge, from "some"of the following:

HUMANITIES: {
    Psychology,
    Psychiatry,
    Philosophy,
    Religion,
    History and
    Management

To Learn **How To manage ourselves**,
First we need the **HUMANITIES**,
To understand ourselves better, to be able:
To modify the Super-Ego*and the Emotional Intelligence,
to Achieve sustainable **success** at:
the Professional, Emotional and Financial level;
thus, "It is for this reason, that **management** will be
more and more, the way through which the:
HUMANITIES
will again acquire worldwide renown,
to produce an impact on business, and
to become an important field
to learn and apply to:
"Practical" business, according to worldwide
famous management adviser **Peter Drucker.**"*(1)

*(1) Management: Tasks, Responsibilities and Practices, by
Peter F. Drucker, Pag. 244, Pub by Harper & Row

To modify our personality, it is advisable,
To improve the super-ego*
to achieve a better
Emotional and Social Intelligence !!!.

*SUPER-EGO:
{Moral Censor}

Habits,	Parents,
Examples,	Relatives,
Beliefs,	Friends,
Values*	Teachers
Teachings	Environment,
Of our:	Religious Advisers,

Super-Ego's Summary =
The **Social** or psychological Environment,
Where we:
Grew, developed and lived and **imitated**
The behaviors and **Values**:
Ethical, Moral, Religious and Cultural.

"Some" of the professional or personal mistakes or failures
could have been avoided, applying a
Modified super-ego*and Emotional Intelligence;
To prevent the appearance of:
Overconfidence or over-complacency,
or arrogance, or hatred, or hostility, or narcissism, or
Not following advice or procedures; leading "frequently" to:
Accidents, mistakes or losses or producing an
Un-conscious drive toward "self-destruction"of the:

a) Career, Business or Marriage and/or
b) The personality or Life.

Consequently, to modify the personality, it is advisable,
To modify the Super-Ego* and the Emotional Intelligence,
To ANALYZE, COMPARE & APPLY
the practical wisdom, given by the world famous professionals,
mentioned in sections I & II of this practical book;
to be able, to achieve sustainable **success** in Life.

Remember how our mind* works:
the thoughts of our mind*
are converted into words,
and those words will become our
Behavior, Values and Habits;
which will define our personality and our:
success in life.

Our Mind*contains our:
Super-Ego = Moral Censor =
Our Social or psychological environment
where we: Grew, lived and developed in life
and this is where we:
Learned, copied and imitated our:
examples or models of
Behavior, Habits and
Values: Ethical, Moral and Religious;
which will define our personality and our:
Success in life.

Consequently, to modify the personality, It is advisable,
"To Be very sincere with yourself,"
To ANALYZE, COMPARE & APPLY
Sections I & II of this practical book,
To be able to follow,
the advice given by world renowned professionals,
To achieve sustainable success at:
The Professional, Emotional and Financial Level.

BE VERY SINCERE WITH YOURSELF.

to speed up the process,
To modify the personality
{Modify the super-ego* and the Emotional Intelligence}.

131

Cont Section III: Executive Summary

"Emotional Intelligence (EI) is a combination of:
1) Self Management and
2) Learning Social skills in an excellent environment,
To transform and optimize,
The performance of the individual or the group."*(10)
The world famous Dr. Emilio Mira y Lopez, MD, mentions:
"Healthy mind in a healthy {body and society}."*(17)pag18
{Achievable Modifying the Super-Ego*and the Emotional Intelligence}.

***Empathy:** It is how well we **Listen** to the point of view of
another person, putting ourselves in their shoes,
to solve problems as a team member;
thereby, implementing an excellent **leadership** role.

EMOTIONS: *in Dr. Goleman's words,*
*"**Personal** competence," comes from being*
Aware of and regulating **One's own** *Emotions.*
*"**Social** competence" is:*
Awareness and Regulation of **Others'** *Emotions."*(8)*

Remember what the:
world famous psychiatrist Dr. Erich Fromm; MD,
Mentions in his book "The Art of Listening," pags.(68,69):
"Only a fundamental transformation of our personality system
can produce a significant change of our character;
in other words:
We must change not only in one aspect,
but in all our aspects of our personality system;
that is, the way we:
Think, act, feel, move .... and everything else,
because, **one isolated emotional change,**
Never produces a lasting effect"
{Advisable To Modify the super-ego* and the Emotional Intelligence}

*(8) Dr. D. Goleman, The Brain and Emotional Intelligence:
New Insights, Published by More than Sound, 2011.
*(10) Special collections, The eading Teams with Emotional Intelligence, by
Drs: Daniel Goleman, R. Boyatsis, A. Mckee, J. R. Katzenbach.
*(17) Dr. Emilio Mira y López, MD: Guia de la salud mental, pag 18, Edit. Oberon.

Dr. Daniel Goleman, a Harvard psychologist, who made
the concept of Emotional Intelligence (EI) popular,
Published recently his new digital book:
The Brain and Emotional Intelligence (EI), New Insights.*(8)

Dr. Daniel Goleman described:
Emotional Intelligence
as having four (4) elements:*(8)

SELF { **AWARENESS**
&
MANAGEMENT

SOCIAL { **AWARENESS**
&
SKILLS *(8)

*(8) Dr. D. Goleman: The Brain and Emotional Intelligence:
New Insights, Published by More than Sound, 2011.
I recommend this important digital book.

*(8) Dr. Daniel Goleman, New Book:
The Brain and Emotional Intelligence, New Insights,
I recommend this important digital book.

**Explanation of Emotional Intelligence and
Its new insights:*(8)**

	SELF	SOCIAL
RECOGNITION	Self Awareness: Emotional Awareness Accurate Self Assessment	Empathy: Organizational Awareness Understanding the Environment
REGULATION	Self Management Self Control: Trustworthiness Conscientiousness Adaptability Drive & Motivation Initiative	Social Skills Influence: Inspirational Leadership Developing others Influence Building bonds Team work and Collaboration

134

*(8) According to Dr. Goleman:
"One of the bad consequences of inadequate
Emotional Intelligence is the Dark side of it; called
The Dark Triad :
The Narcissist, Machiavellians and Sociopaths,*(8)
Because they **lack** Emotional **empathy**; thus,
These people are sociopaths, and they
Do not care about the human consequences of
Their lies or manipulations, and they
Do not have regrets about inflicting cruelty."*(8)

"The sociopath's brain has deficits in several
Areas key to Emotional Intelligence:
The anterior cingulate, the orbitofrontal cortex,
The amygdala and insula, and in the connectivity of
Those regions to other parts of the brain."*(8)

"Some of the consequences of these sociopaths are
Recognizable in organizational life:
  1) The Bullying
  2) The "Kiss-Up-Kick-Down" Boss
     Who can be very charming to superiors,
     But abusive and tyrant to subordinates.
     Another is the embezzler, a crook.
  3) The freeloader, the person who always holds a
     coffee cup and never does a lick of work."*(8)

*(8) Dr. Daniel Goleman:
The Brain and Emotional Intelligence, New Insights,
Published by More than Sound, 2011.

The advisable School and University formation:
The world famous psychiatrist:
Dr. Emilio Mira y López; MD,
Mentions in his book "Guide to Mental Health":
That to achieve a modified personality,
"The school or University, must place more emphasis on
The "formation" than on the "instruction" of its students; **b**
Teaching them, from the early years,
How to solve, with calmness and decision,
The difficult situations,
They will encounter in real life and in theirs personal lives."

"Dr. Emilio Mira y Lopez, MD, suggests to follow two (2) steps,
To achieve the proper **educational "formation"**:
1) Create in each student a feeling of
Inner peace and trust in oneself.
2) Avoid conflict between
Real ambitions and realizations,
Regarding wishes and successes,
In other words, achieving what was intended originally,
To avoid falling prey to "imaginary" achievements;
That may result later in addictions to:
Gambling, drinking, imaginary fictions, or neurosis."

"To achieve this mission, it is necessary,
The collaboration between school and family;
and the teachers, must work together
With the students' family to convince them that:
A person is worth not for what he knows and feels,
But for what "good" he is capable to do,
With what he knows and feels."

"The true educational role, of the school or university is
To make people understand, that the real worthiness of a person
Is to be **the owner of his own destiny,** and to know:
How to procure first, the wellbeing to everybody, before his own;
That is, the real educational formation of the school {or University},
To collaborate in the fight against the present social diseases and vices."

The MAL-INTENTIONED "Humorist"

The psychiatrist: Dr. Emilio Mira y Lopez; MD,
Describes the mal-intentioned Humorist,
as an **angry** failure with inner fear, and
"sometimes" resorts to bad {jokes or humor},
To be able to say it humorously,
what they are not able to say it seriously.

In this case, the bad {jokes or fun} made at the
expense of somebody else, is made by a person
who is resented, who has inner **anger;**
showing how much he despises himself, and
he feels the same as the old court joker.

To protect yourself psychologically from the
Mal-intentioned Humorist,
You must start urgently the process
To Modify your personality**
{Modify the super-ego* and the Emotional Intelligence},
To prevent the Mal-intentioned humorist
From impairing your **success** at the:
Professional, Emotional and Financial level.

**ANALYZE, COMPARE & APPLY
The advice of the world renowned professionals
Mentioned in Sections I & II of this practical book,
To be able, to achieve professional and emotional success.

Warning: This book is not intended to take the place of medical advice
From a trained medical professional; thus, always seek the best medical
Assistance in ALL cases regarding your total health. Neither the authors,
Nor the publisher of this book, nor the publishers of the books mentioned
in this book take any responsibility if this warning is not followed completely.

Cont Section III: Executive Summary

I will mention a brief story, that will help us
To Become   Humble and
The story is the following:
It happens at "Oxford" University,
a "newly" graduate told an old professor:
Thank you very much professor:
with the education that I have received,
I feel that I am completely prepared for life **!!**;
and the old professor answered him very humbly:
In my case my son:
Only until now I feel completely prepared for life.

Perhaps, this is a good time to mention, that even
if one has two (2) University degrees, the maximum
that they can teach us at the University is around
twenty **(20%)** percent,
The other **80%** must be obtained throughout:
Life's practical experiences, attending:
Seminars (Locals, Nationals, & Internationals),
consulting with your colleagues,
Studying (keeping up-to date),
attending post-graduate school,
And doing your own R & D*;
In other words: there is plenty to learn, even
after we obtained the University degrees.

It will be quite appropriate to mention,
The words of wisdom of the world famous cardiologists:
Drs. Friedman, Rosenman, MDs, Ulmer & Associates:
**"We modern physicians and nurses
still have much to learn, and
much to be modest about."***(4) pags, 120-1

This excellent advice was given, to remind us,
to keep an open mind (flexible super-ego),*
For the continuous new developments
Our respective fields of work.

The book: "Snakes in Suits: When Psychopaths Go to Work" by
 Drs. R. D. Hare and P. Babiak, Phds.*(7), mentions that
 Some of the red flags of the psychopaths might be:

1) Superficially charming,
2) Prepotent, or arrogant, or narcissists,
3) Lie & manipulate people easily,
4) Lack remorse, empathy,
5) Are cold, in-considerate and
      Mean, and do not accept responsibility.
6) Only care about: Money & Position,
7) Are irritables, impatients & impulsives: (the 3 I)
8) Are Bullies/or abusive with their subordinates,
      But very nice to their superiors,
      Thus, creating: Low morale or
      Loss of: Productivity or satisfaction or
            Market share & Earnings ,
      and also high personnel turnover *(7)

"Dr. S. Freud tried unsuccessfully to explain."*(8)
      Why the psychopath has not yet resolved,
Some very important unconscious childhood's issues
{Relating to the unresolved unconscious hostility toward one of the parents}.

According to Dr. S. Freud,
      The child UN-consciously hates one of the parents;
Thus, lacks that important parental censorship activity, and
      Continues braking all the norms or duties taught,
      Lacking - according to Dr. S. Freud -
      The so called Super-Ego* {or Moral Censor}"
{All this information, is mentioned by the world famous psychiatrist,
      Dr. Emilio Mira y Lopez, MD, in his book,
      "The four Giants of the Soul"
"The Fear, Anger, Love and Duty", pag. 201-2, publ. by Lidiun}.

"Sometimes the Psychopath has Brain's abnormalities in the sections
of the Ventrolateral, orbitofrontal Cortex and Amygdala."*(8)

*(8)The Brain and Emotional Intelligence: New Insights, By Harvard's Psychologist,
Dr. Daniel Goleman,Published by More than Sound, 2011. I recommend this Important digital book.

According to the ex-CEO of G.E. Jack & Suzy Welch,
They mention the following in theirs book
The Real Life MBA*(19):

1)   "Real **Leadership** is based on **2 Ts: Truth & Trust.**
The **Leader** also utilizes the **4 E: Energy, Energize, Execute & Edge;**
To have the courage & passion to make tough and risky decisions,"*(19)
Applying an excellent modified Super-ego* and Emotional Intelligence.

"A Leader also has excellent **discernment,**
To understand, appreciate and respect,
To analyze situations, problems or decisions"*(19)
{Excellent Modified Emotional Intelligence and super-ego*}.

"Table: Typical Evaluation of Employee: Performance and Behavior:"*(19)

EXCELLENT / GOOD	NEEDS IMPROVEMENT
A) 20% = Are super stars	C) 10% = Low performance
B) 70% = Are average	*(19)

*(19)The Real life MBA, by Jack & Suzy Welch, pags 67,125,151, Pub. Harper, NY.,NY.

Cont. According to the ex-CEO of G.E. Jack & Suzy Welch,
They mention the following in theirs book
The Real Life MBA*(19):

2) **"The Innovation** = Is the integral sum of small innovations = Σ = $\int$,
The Innovation = Σ = $\int_{n=1}^{n=infinite}$ = Innovations =Σ from n=1, to n=infinite."
"Daily Innovation = Everybody looking for better ways to do things;
To achieve the best product or service, with
The best: quality, safety, and price."*(19)

3) **"Strategic planning:** Flexible and agile, because the market conditions
Move and change all the time: Locally, nationally and globally.
To apply the best practices and obtain a competitive advantage,
It is advisable, to select the best personnel, with the best theoretical
and practical knowledge, and assign them to the most di cult jobs;
To be able to optimize the Innovation and competitiveness:
Locally, nationally and globally."*(19)

4) **"Team work: Always allow every member,**
To express theirs solutions or ideas
To improve the innovation and competitiveness;
This strategy will help, to give dignity to the team members; thus,
They will feel part of the team, corporation, contractor or profession."*(19)

5) **"** The brilliant ex-CEO of G. E., Jack Welch mentions that:
"A single person, with a bad temper
Can destroy, the work of a whole department, and
Can infect the whole corporation, contractor or profession."*(19)
{It is advisable, to Modify the Emotional Intelligence and the super-ego*}

Summary: "It is advisable, to create a pleasant working environment,
To make everybody feel appreciated and recognized,
For theirs performance and behavior; thus,
Improving productivity and personal satisfaction."*(19)

*(19)The Real life MBA, by Jack & Suzy Welch, pags 67,125,151, Pub. Harper, NY.,NY.

According to the world famous, Dr. Luis Rojas-Marcos, MD,*(17):
Innovation and Competitiveness:
"One way To compete
(In a noble and decent way)
Is creating an
Environment:

Positive, stimulating and
Creative environment,"*(17) with a
Modified: Super-ego* and
Emotional Intelligence.

"The Innovation,
Allow us
To optimize:

The relations,
The work and
The time to:

Improve ourselves,
Advance, and
Compete better.

Also, when analyzing Companies or projects,
Always remember Murphy's famous law:
If there is a chance that something will go wrong, it will,
Unless we take precautions to avoid or mitigate it; also
If there are 10 good positive outcomes, and 2 negative outcomes,
Frequently **the 2 negative outcomes will happen first.**
Thus, when performing the contingency plans, always
Utilize an excellent modified Super-ego*and Emotional Intelligence
For the three (3) case scenarios: the best, medium or worst.

Note: HOW **To Delegate** a Task or a Goal:
We can delegate a task or a goal to be achieved; **but**
We can not delegate the responsibility of controlling:
That the task or goal is being achieved with the **best:**
Quality, safety, on time and under budget; **thus,**
The Controlling responsibility always belongs to the Manager & the team.
Utilize an excellent modified Super-ego*and Emotional Intelligence.

*(17) Dr. Luis Rojas-Marcos, MD: "Todo lo que he Aprendido" pag 132
Pub. Espasa, Spain, 2014

Always remember the golden rule of
Psychology: "Be very sincere with yourself"
To be able, to obtain an honest
Internal Self-Evaluation:

1) "Determine how Intelligent and creative at your job.
Examine your sense of humor and flexibility:
To Modify your personality (the super-ego)."*(4)

2) "Examine all the "activities" that now absorb your:
Intellectual, Emotional, and Spiritual interests.
How many of these activities have to do with
With your concern for: Art, literature, music,
Drama, philosophy history, science??."*(4)

3) "Seek and assess your free-floating hostility."*(4)

4) "Estimate, the ease to receive and give:
Loyalty, and affection."*(4)

5) "Determine the courage and fear you possess."*(4)

6) "Examine your ethical and moral principles.
How honest have I been in my life ?."*(4)

7) "Finally, ask yourself:
What should be the essence of my Life ?."*(4)

*(4) Treating Type "A" Behavior & Your Heart,
Pgs.(95,218,219,242) by: Meyer Friedman, MD &
D. Ulmer, RN, MS, Published by Fawcett Crest, NY.

Harvard's psychologist,
Dr. Daniel Goleman, mentions in his book,
"The Brain and Emotional Intelligence; New Insights"*(8):
To Modify "Old" Habits,
"It takes (3-6) months of Practice,
Before the **"New"** habit comes more naturally than the old."*(8)

Consequently, to modify the personality, It is advisable,
"To Be very sincere with yourself,"
To ANALYZE, COMPARE & APPLY
Sections I & II of this practical book,
To be able to follow,
The advice given by world renowned professionals,
To achieve sustainable success at the:
Professional, Emotional and Financial Level.

BE VERY SINCERE WITH YOURSELF.

To obtain the soft skills required, to improve
The Emotional + Social Intelligence,
To be able, to listen with excellent **Empathy,***
and modify also the super-ego;* allowing us,
"To Improve or regain **40% of the:**
Quality, safety, performance, leadership,
Innovation, competitiveness and earnings of any
Corporation, contractor or profession
{Mr. H. Geneen, Ex-CEO of ITT}"*(2)
{Advisable to modify the Super-Ego* and the Emotional Intelligence (E.I.)}.

***Empathy:**
It is How well we Listen,
To the point of view of another person, by
Putting ourselves in their shoes, To solve
Problems as a team member; thereby,
Implementing an excellent Leadership role.

Definition of "Wealth":

There are two (2) components of Wealth:
1) **Inner** = Happiness = Satisfaction with life
   Modified Super-Ego*and Emotional Intelligence
   Excellent Personality (Character)
   Inner peace, harmony = **60%**

   } **Wealth**

2) **Outer** = Assets, Positions,
   Titles, Diplomas & money ($$$) = **40%**
                                    100%

Summary: Improving our **Inner** wealth,
Increases or preserves our **Outer** wealth.
Wealth is composed of two (2) elements:
1) The **Inner Wealth = 60%** of our total wealth,
   It gives us Happiness = Satisfaction with life, inner peace,
   to be able, to apply the golden rule of psychology:
   "Be very sincere with yourself,"
To modify the Super-Ego,*and the Emotional Intelligence,
To obtain an excellent personality (Character), to be able:
   a) To avoid mistakes, accidents, or losses, and
   b) Improve our **Inner** wealth; thus,
      Increasing or preserving our **Outer** wealth, and also our:
         Mental, physical and financial Health.

2) The **Outer Wealth = 40%** of our total wealth,
   Which is always exposed to the mortal trap, mentioned
   By the ex-President of ITT, H. Geneen: sometimes,
"More fortunes (or careers) **are destroyed by success than by failure,**
   Frequently, generating Egotism"**(2)
= Arrogant, abusive or disrespectful behavior; sometimes,
   Causing harm to others or themselves.

Conclusion: If we utilize our improved **Inner** wealth,
To modify the super-ego* and the Emotional Intelligence; then,
We will be able, to increase or preserve our **outer** wealth, and also
Succeed at the emotional, professional and financial level.

According to the world famous, Dr. Emilio Mira y Lopez, MD:
*The super-ego = Anti-ego, because there is a sector, of psychological forces
Hostile to the ego. The super-ego is the psychological or **Social** environment,
Where we: Grew, lived, studied and **Imitated** the Behaviors and
Values: Ethical, moral, religious and cultural.

**(2) Managing, by Harold Geneen (Ex-CEO ITT), Pag (184-5) Pub. by Avon Books.

145

Practical Applications of Good Emotional Intelligence:
PERSONAL and PROFESSIONAL MANAGEMENT:

Please always remember, the Mind* contains our:
Super-Ego = Moral censor, or
Psychological **or Social environment,** where we:
Grew, lived and developed in life;
and this is where we:
Learned, Copied and Imitated
Our examples or models of:
Behaviors, Habits and
Values: Ethical, Moral and Religious;
Which will define our personality and
Our success in life.

"Some" of the professional or financial mistakes or failures,
Could have been prevented or avoided
If those persons had:
Modified the super-ego*and Emotional Intelligence;
Thus, being able to avoid the:
Overconfidence, or over-complacency, or arrogance,
or hatred, or hostility, or narcissism, or
not following advice or procedures;
Leading "frequently" to major:
Mistakes, or losses or causing an
Un-conscious drive toward "self-destruction" of the:
a) Career or Business or Marriage and/or
b) The personality or the **Life.**

Consequently, it is advisable,
To acquire more insight into our own:
Strengths and Weaknesses,
To be able to modify our personalities;
To obtain sustainable success at:
The Professional, Emotional and Financial level
{It is advisable, to Modify the super-ego* and the Emotional Intelligence}.

The brilliant ex-Director of ITT, Mr. Harold Geneen,
Mentions in his book: Managing (Pag. 185)*(2), that:
If we eradicate the "Egotism"** in "some" companies,
{Due to un-modified super-ego* and Emotional Intelligence}
We can:

Improve { Performance, } of any
40% { Productivity & } Company or
in: { Earnings } Contractor"*(2)

This is one of the reasons,
Why it is very important for professionals
To Improve **"How to manage Themselves"**, to obtain:

a) An excellent {Emotional + Social} Intelligence
b) To cultivate the: Inner {Peace or Tranquility}
c) To be able, to apply:
The golden rule of psychology:
"Be very sincere with yourself,"
To modify the super-ego*and the Emotional Intelligence,
d) To achieve:
Sustainable success at:
The Professional, Emotional and Financial Level.

Mr. Geneen also mentions a very wise advice:
"More Careers are ruined by Success than by Failure"
{Due to the un-modified super-ego*, sometimes their
Success goes way over their heads, and
Frequently, they start mistreating others or themselves}.

It is advisable, to Modify the super-ego* and the Emotional Intelligence
"To Be very sincere with yourself," when
Reviewing sections I & II of this practical book, to be able to follow,
The advice given by world renowned professionals,
To continue achieving sustainable **success** at:
The Professional, Emotional and Financial Level.

*(2) Managing, By: Harold Geneen, Pag185, Pub Avon Books.
*Egotism = Arrogant, or abusive or disrespectful Behaviors.

Matching points (*=*) of the **Management styles** of:
*Segundo Cristancho J. & Company {during 67 years}, and
*"Mr. Harold Geneen, Ex-CEO of ITT
{Helped a **$766** million telephone company grow into a
$22 Billion multinational conglomerate, according to
His book: Managing, Pub. by Avon Books}:*(2)
Several executives, that worked with Mr. Geneen,
graduated to successful careers as Fortune 500 CEOs:
*=*1) **Plan, direct and control** "Know every facet of the businesses" and
= 2) Treat people with respect & admiration and listen with empathy,
Have an open door, but avoid incompetence and neglect.
*=*3) Urgently, If present, modify the egotism,**
To be able, "To improve or regain 40% of the:
Quality, safety, performance, productivity,
Innovation, competitiveness and earnings of any
Corporation, contractor or profession
{Mr. H. Geneen, Ex-CEO of ITT}."*(2)".
{Modify the super-ego* and the Emotional Intelligence,
Applying sections I & II of our practical book}.
*=*4) "Update the plans, associates must update them too.
Use what worked, but changed when necessary."
*=*5) "Be a: {Hands on, present, active and a responsible} manager."
*=*6) "Know:{your business, what works, your people, who works}
*=*7) "Manage with respect & admiration (w/o egotism)."**
"Performance and results must be achieved,
If you don't achieve those results, you're not managing"
{To Improve your management, apply sections I & II of our book}.
*=*8) "Put in the time and dedication to work things out."
Never delegate the responsibilty of controlling a
Task or a goal, to be done, with excellent:
Quality, safety, on time and under budget.
*=*9) "When a client reports a complaint or suggestion,
Communicate it to management ASAP, and fix it asap,
According to the plans, contract and specifications,
To comply with the client wishes and desires; thus
always having a satisfied and loyal client."

*Also, applied at: **Segundo Cristancho J. & Co.** during 67 years.
*(2)Managing: Harold Geneen, ex- CEO, ITT, Pub. by Avon Books.
**Egotism = Arrogant, or abusive or disrespectful Behavior

Remember the following advices:
HOW To DELEGATE a Task or a Goal:
We can delegate a task or a goal to be achieved;
But we can not delegate, the "permanent" responsibility of controlling:
That the task or goal be achieved with the **best:**
Quality, safety, on time and under budget,
That **"permanent" responsibility of controlling**
Always pertains to the contractors, professionals and the team.

It is advisable to follow the advice of Mr. H. Geneen (Ex-CEO-ITT),
If present, we must eradicate the Egotism;
{Modify the super-ego* and the Emotional Intelligence}
Applying sections I & II of our practical book};

"To be able to:

$$\text{Improve } 40\% \text{ in:} \left\{ \begin{array}{l} \text{Performance,} \\ \text{Productivity \&} \\ \text{Earnings (\$\$\$)} \end{array} \right\} \text{of any Company"*(2)}$$

*Super-Ego = Environmental stimulus, where we:
Grew, lived, imitated and developed.

|Please, do not borrow money **"above"** your leverage ratios of:
30% < Your Liquid Assets < 50%.
Watch your **financial ratios,** especially the acid test;
UOPM= Using other People's Money has:
(+/-) Positive and negative consequences = Bankruptcy.
To act wisely in your financial management,
Modify the super-ego*and the Emotional Intelligence.

Leadership & Competitiveness can be:
Learned, imitated and improved greatly, **if** the
Super-ego*and the Emotional Intelligence have been modified previously.

Quality, Safety, Innovation can be improved when:
Our mind is fully focus and engaged, with no distractions, with
The super-ego*and Emotional Intelligence modified.

The fight against BAD TEMPER:

It is advisable, to Modify the super-ego* and the Emotional Intelligence,
According to world famous, Dr. Emilio Mira y López, MD,
In his excellent book "The four big Giants of the Soul",
"Fear, Anger, Love and Duty"*(5), he mentions the following:

"It is convenient to remember that, anger comes from fear,
and anger blinds the sight and the understanding; thus,
If we want to avoid being a victim of that angry impulse,
Then we have to start"*(5)
———— "**Knowing ourselves better** ———— because the more
We know about ourselves, the easier it will be to identify:
1) The good natural talents to be developed, and
2) The personality's deficiencies that will be convenient to eliminate,
To be able to modify the Super-ego*and the Emotional Intelligence;"*(19)
"Thus, the afflicted person can stop suffering, and
That person will stop making other people suffer."*(5)

"The bad temper is a sign of deep anger,
Showing signs of insecurity, lack of self control, and
Lack of faith in his own capabilities."*(5) pags.124,5

"Be careful, the bad temper is contagious,
A single person with bad temper can infect the whole group;"*(5)
The brilliant ex-CEO of G. E., Jack Welch mentions that:
"A single person with a bad temper
Can destroy the work of a whole department, and
Can infect the whole corporation or profession."*(19)

Summary: "It is advisable, to know ourselves better, to be able,
To modify the Super-ego* and the Emotional Intelligence,
To diminish or avoid the internal anger; allowing us,
To live in peace with ourselves and others."*(5)

*(5)Dr. Emilio Mira y López, MD: "The four big Giants of the Soul",
    "Fear, Anger, Love and Duty", pags.124,5, Publ. Lidiun.
*(19) The Real MBA, by Jack & Suzy Welch, pub. by Harper, N.Y., N.Y.

Definition of "Wealth":

There are two (2) components of Wealth:
1) **Inner** = Happiness = Satisfaction with life
   Modified Super-Ego*and Emotional Intelligence
   Excellent Personality (Character)
   $\qquad$ Inner peace, harmony = **60%**

2) **Outer** = Assets, Positions,
   Titles, Diplomas & money ($$$) = **40%**
   $\qquad\qquad\qquad\qquad$ **100%**

} **Wealth**

$\qquad$ Summary: Improving our **Inner** wealth,
$\qquad$ Increases or preserves our **Outer** wealth.
Wealth is composed of two (2) elements:
1) The **Inner Wealth** = **60%** of our total wealth,
   It gives us Happiness = Satisfaction with life, inner peace,
   To be able, to apply the golden rule of psychology:
   $\qquad$ **"Be very sincere with yourself,"**
To modify the Super-Ego,*and the Emotional Intelligence,
To obtain an excellent personality (Character), to be able:
   a) To avoid mistakes, accidents, or losses, and
   b) Improve our **Inner** wealth, thus,
   $\qquad$ Increasing or preserving our **Outer** wealth, and also our:
   $\qquad$ Mental, physical and financial Health.

2) The **Outer Wealth** = **40%** of our total wealth,
   Which is always exposed to the mortal trap, mentioned
   By the ex-President of ITT, H. Geneen: sometimes,
"More fortunes (or careers) **are destroyed by success than by failure,**
   Frequently, generating Egotism"**(2)
= Arrogant, abusive or disrespectful behavior; sometimes,
   Causing harm to others or themselves.

$\qquad$ Conclusion: If we utilize our improved **Inner** wealth,
To modify the Super-Ego* and the Emotional Intelligence; then,
We will be able, to increase or preserve our **outer** wealth, and also
Succeed at the emotional, professional and financial level.

According to the world famous, Dr. Emilio Mira y Lopez, MD:
*The super-ego = Anti-ego, because there is a sector, of psychological forces
Hostile to the ego. The super-ego* is the psychological or **Social environment,**
Where we: Grew, lived, studied and **Imitated** the Behaviors and
Values: Ethical, moral, religious and cultural.

**(2) Managing, by Harold Geneen (Ex-CEO ITT), Pag (184-5) Pub. by Avon Books.

The book: The Brain and Emotional Intelligence, New Insights,
By Harvard's psychologist, Dr. Daniel Goleman,
I recommend this important digital book.*

According to Dr. Goleman, **to modify "old" habits,**
"It takes (3-6) months of practice, before the
"New" habit comes more naturally than the old."*(8)

Description of the Social Emotional Learning (SEL) programs:

Dr. Goleman mentions that, at some schools in the USA,
New programs called **SEL: Social Emotional Learning**
Have been implemented,
To improve the Emotional Intelligence of students."

"On the wall of every classroom there is a picture of a
Stoplight, with its red, yellow and green lights."*(8)

1) RED light means:
STOP = **Calm down, and think before you act;**
Teaching students, that:
You can not control what you are going to feel,
But you can decide what to do next.

2) YELLOW light means:
Think of the range of things you might do,
and what their consequences would be; thus,
Enabling the student to learn to pick the best alternative.

3) The **GREEN** light means =
Try it out, and see what happens,
This is drilled into the students, and it works;"*(8)
Thus, "achieving the main purpose of learning
How to cope with disruptive feelings, to be able
To get along better with other people, and
Improve the performance at:
School or work."*(8)

*(8) Dr. D.Goleman: The Brain and Emotional Intelligence:
New Insights, Published by More than Sound, 2011.

This definition applies to all the professions pertaining to:
Engineering & Construction:
Contractors, Engineers, Architects, Inspectors and Unions.

Definition of a **PROFESSIONAL:**

1) A true professional possesses two types of knowledge:

Types of knowledge
{
1) The Professional or practical wisdom, and
2) The **one self's knowledge** =
Modified: {Emotional Intelligence and Super-Ego;*
}

To avoid mistakes, accidents or losses.

2)   Always applies the golden rule of psychology:
"Be very sincere with yourself,"
To modify the Emotional Intelligence and the Super-Ego* pertaining to:
"The Fear, Love, Duty and
Anger: Hatred, enviousness, prepotency, bad temper, and egotism,**
Mentioned by the world famous Dr. Emilio Mira y Lopez, MD
{In his book, "The four Giants of the Soul", Edit. Lidiun}."*(5)

3) Optimizes the relationships with others, to listen to them with
Empathy, respect and admiration; to implement the
Innovations and changes required,
To improve or recuperate 40% of the
Quality, safety, performance, leadership,
Innovation, competitiveness and earnings of the:
Country, corporation, contractor, project or profession.

4) A Professional continuously improves the Inner wealth;
To be able, to improve or preserve the **Outer** wealth.

On Nov. 1/202, Business Week mentioned that,
"Some Professionals are:
Hired for their excellent professional knowledge and
Fired for their Personalities" = Un-modified Super-Ego*and Emotional Intelligence.

According to the world famous,
Dr. Emilio Mira y Lopez, MD, the
*Super-ego = Anti-ego, because there is a sector, of
Psychological forces hostile to the Ego.*(5) pag202

The super-ego is also, the psychological or **Social environment,**
Where we: Grew, lived, studied and **Imitated** the behaviors and
Values: Ethical, moral, religious and cultural.
**Egotism = Arrogant, or abusive or disrespectful behaviors.
*(2) Managing, by H. Geneen, Ex-CEO ITT, Pag (184-5) Pub. by Avon Books.

How To Achieve a lasting Personality {Character} "Modification":
In this book, we explained how the human mind behaves, and we
Described **The Structural components of The "Personality"** as follows:

1) The ID   2) The EGO   3) The SUPER-EGO

*The ID: { Instincts, Impulses, Passions } Irrationals & Un-Conscious

Controls the Pain & Pleasure

The EGO: { Perception of: Reality & Reason

Tries To Regulate & Control the ID

SUPER-EGO: { Habits, Examples, Beliefs, Teachings, Behaviors, *Values, of Our: } Parents, Relatives, Teachers, Neighborhood, Religious Advisers, Friends

{Moral Censor}

Psychological or Social Environment
where we: grew, lived, studied and
Imitated the Behaviors & Values:
*Ethical, moral, religious and cultural.

According to the world famous psychiatrist,
Dr. Emilio Mira y Lopez, MD;
*"The super-ego should be called Anti-ego, because sometimes,
There is a sector, of psychological forces hostile to the Ego."*(5)pag202

The super-ego also defines, the psychological or **Social environment,**
Where we: Grew, lived, studied and **Imitated,** the
Behaviors and **values:** Ethical, moral, religious and cultural.
*Super-ego = In psychoanalysis, is that part of the psyche which:
   1) Is critical of the self or Ego, and
   2) Enforces moral standards, at an unconscious level,
      {**If** we have modified all the deficiencies of the Super-Ego};
      Then, it blocks the irrational unacceptable impulses of the Id.*

Summary: Hypothesis
How To Achieve a lasting Personality {Character} "Modification":
The super-ego = In psychoanalysis, is that part of the psyche which:
1) Is critical of the self or Ego, and
2) Enforces moral standards, at an unconscious level,
    {**If** we have modified all the deficiencies of the Super-Ego};
    Then, the Super-Ego **blocks** the irrational unacceptable impulses of the **Id.***

*The **ID:** Instincts, Impulses, Passions } Irrationals & Un-Conscious
Controls the (Pain & Pleasure)

To enable, our super-ego* to block the irrational:
Impulses, instincts or passions of the Id,*
It is advisable, to modify all the learned and imitated
Deficient behaviors and values of the super-ego pertaining to:
The Fear, Love, Duty and
Anger: Hatred, enviousness, prepotency, bad temper, and
Egotism** = Arrogant, or abusive or disrespectful behaviors.

Also it is advisable to modify -if present-:
The hyper-aggressiveness, hurry-sickness, mal-intentioned humor;
And lastly -if present-,
The un-conscious self-destruction of the:
Career, business, marriage or **life**
{Mentioned by Dr. M. Friedman, MD, and also by the world famous,
Dr. Emilio Mira y Lopez, MD; in his book: "The Four Giants of the Soul"}.

Thus, we must continue improving our **Inner** wealth & health,
To increase or retain our **Outer** wealth & health;
To be able, **"To improve or regain 40%** of our:
Performance, productivity, leadership and earnings of any
Corporation, contractor or profession."
{According to H. Geneen, ex-CEO of ITT}.*(2)

Thus, to achieve a lasting personality modification, it is advisable to apply,
The advice given in the book "The Art of Listening, Pags.(68,69), by
The world famous Dr. Erich Fromm, MD:
"Only a fundamental transformation of our personality system,
Can produce a significant change of our character; that is,
The way we: Think, Act, Feel, Move .... and everything else,
Because, **one isolated Emotional change,**
Never produces a lasting effect"
{Therefore, it is advisable, to modify the super-ego* and the Emotional Intelligence}.

HOW to Improve your professional SALES or SERVICES:
To implement the following sales' recommendations,
It is advisable first, to modify
The super-ego* = Anti-ego;* such as:
Anger, or prepotency, or enviousness, or bad temper or egotism.**

The following are the recommended keys,
To obtain or retain clients and
Improve your professional SALES or SERVICES:
1) **Enthusiasm,** to obtain or retain prospective clients:
Always, visit them personally, and keep in touch regularly
Showing the client how your products or services can be
The best solution to the client's needs and wants.
2) **Listen** with improved:
Empathy, respect, admiration and Emotional Intelligence;
To be able, to learn what are the client's needs and wants,
and always, looking out, for the best interest of the client.
3) Always have an excellent product or professional service,
Offering the best: quality and safety,
At the best price possible
{Without sacrificing safety or quality},
4) Excellent Knowledge of your:
a) Product or Profession and
b) Professional Association or industry as a whole, and
5) Always act with the best personal
Honesty, integrity and dependability;
After all, people buy
Services or products from people

According to the world famous,
Dr. Emilio Mira y Lopez, MD:
*"The super-ego should be called Anti-ego, because sometimes,
There is a sector of psychological forces hostile to the Ego."*(5)pag202

*The super-ego also defines the psychological or **Social environment,**
Where we: Grew, lived, studied and **Imitated** the
Behaviors and **Values:** ethical, moral, religious and cultural.

**Egotism = Arrogant, or abusive or disrespectful Behaviors.

1) Remember President Kennedy's famous phrase:
"Do not ask what your country can do for you;
ASK what You Can Do for your country."
If the present and future Professionals really want
To help create overall wellbeing; then, it is advisable
To remember that, the real purpose of a corporation or
Profession is to promote the good and wellbeing of the:
Stockholders, partners, employees, unions and the global society.

2) "If the present and future professionals
Really want to become deeply involved in the:
Strategic management of the corporation, contractor or profession;"
Then, they must become more versed in many
Other areas of expertise well beyond finance, such as:
Modifying the Emotional Intelligence and the super-ego,*
And that sometimes, takes a few months of:
Learning, practice, cooperation and patience.

To modify the Super-Ego*and the Emotional Intelligence,
It is advisable, to apply the golden rule of psychology:
"Be very Sincere with Ourselves," because the more
We know about ourselves, the easier it will be to identify:
1) The good natural talents to be developed, and
2) The personality's deficiencies that will be convenient to eliminate;
To Improve or Recuperate 40% of the:
Quality, safety, performance, leadership,
Innovation, competitiveness and earnings of the:
Contractors, Engineers, Architects, Inspectors and Unions.

Consequently, dear Professionals:
The world's destiny is up to you,
The future of mankind is in your: Hands, Heads & Souls !!!
Respectfully yours.
Ing. Emilio Cristancho-G.: B.S..M.B.A
According to the world famous psychiatrist,
Dr. Emilio Mira y Lopez, MD:
*"The Super-Ego should be called **Anti-Ego,** because
There is a sector, of psychological forces hostile to the Ego."*(5)pag202
The Super-Ego also defines, the psychological or **Social environment**
Where we: Grew, lived, studied and **Imitated,** the
Behaviors and Values: Ethical, moral, religious and cultural.
**Egotism = Arrogant, or abusive or disrespectful Behavior.

BOOK'S SUMMARY:

Contractors, Engineers, Architects, Inspectors & Unions
"Can Improve or Recuperate 40% of the:
Quality, safety, performance, leadership,
Innovation, competitiveness and earnings of any
Corporation, Contractor or profession."*(2)

New definitions of:
A Professional and also Wealth;
To help us obtain sustainable success:
Professionally, Emotionally and Financially.

If we improve our Interior wealth; then
We can increase or preserve our Exterior wealth.

How to modify the personality {Character},
To be able to identify:
1) The good natural talents to be developed, and
2) The personality's deficiencies that will be
Convenient to eliminate.

Two types of knowledge are required,
To become an excellent professional;
To be able, to improve the processes of
Innovation, competitiveness, leadership and
Earnings of the:
Contractors, Engineers, Architects, Inspectors & Unions.